Visions

To Martin

Thanks for mentoring me!

All the best

23/11/18

Hispanic Studies: Culture and Ideas

Volume **6**

Edited by
Claudio Canaparo

PETER LANG
Oxford · Bern · Berlin · Bruxelles · Frankfurt am Main · New York · Wien

Sergio Plata

Visions of Applied Mathematics

Strategy and Knowledge

PETER LANG
Oxford · Bern · Berlin · Bruxelles · Frankfurt am Main · New York · Wien

Bibliographic information published by Die Deutsche Bibliothek
Die Deutsche Bibliothek lists this publication in the Deutsche Nationalbibliografie; detailed bibliographic data is available on the Internet at ‹http://dnb.ddb.de›.

British Library and Library of Congress Cataloguing-in-Publication Data:
A catalogue record for this book is available from The British Library, Great Britain, and from The Library of Congress, USA

The graphic design of this book was made by Florian Ziche.
The illustration of the cover is a reproduction from a work entitled 'Uriarte y Blas Parera' from the Argentinean plastic artist Claudia del Rio.

ISSN 1661-4720
ISBN 978-3-03910-923-4

Hochfeldstrasse 32, Postfach 746, CH-3000 Bern 9, Switzerland
info@peterlang.com, www.peterlang.com, www.peterlang.net

Printed in Germany

To Emily

Contents

Acknowledgements

I would like to thank greatly to two persons, without whom the completion of this book would not have been possible, not only for their intellectual uphold, but also for their personal aid: firstly to Prof. Eduardo Ortiz for his support in hard times, thorough comprehension and for his superb soul which he gives to everything he does; and dearly to Dr. Claudio Canaparo for his marvelous encouragement, unconditional friendship and his faith in all daring intellectual projects.

I would like to thank also to the Fereday family for everything and people abroad who helped me all the way my project, specially to Conchita Martínez, Lula Carbajal, Dr. Alejandro Garciadiego and Dr. Ricardo Berlanga,

London
November 2005

Presentation

This work is an attempt to make a multidisciplinary study on Applied Mathematics. The leading cases are 1) Operational Research, an branch of mathematics which had its origins in Britain and which is the token *par excellence* of the modern application of mathematics in the 'real world', 2) Statistics, a branch which took a major role in our daily lives since it incorporated the modern concept of information, and 3) Applications of Functional Analysis another branch of modern mathematics which involves direct applications and breakthrough models for computational interfaces.

These areas of mathematics help to support and influence executive decisions at any level in many organizations from their roots, affecting many spheres of modern life.

By making use of historical, archival and textual sources in an epistemic, semiotic and hermeneutic ground, I begin with the analysis of one of the most strategic and dynamic of all industries: *the militaries*. Of course the analysis covers mainly the activities in war, however it is also contemplated the peace times so the role of science is grasped to the maximum sense not only in their technical parts but also in the sociological, institutional, political, economic, military and epistemological. Then a continue with another important area in any modern institution: *research and development*, where I discuss also topics applied to a Latin American case, through mainly the work of top scientists whose contributions cover a wide angle of knowledge. Finally I analyze an economic case for which even a Nobel Prize was awarded, and for which decisions in many modern corporations are based.

Although this is a book for which the main topic Mathematics I intended to write it aiming a wide variety of audiences, above all those interested in Latin American Cases, as it focuses on it in the application of the theories developed here; many people will be able to read it as it presents many aspects of the mathematical activity unveiling, organizational, administrative, psychological and mainly linguistic and philosophical sides of Mathematics in a tight and rigorous historiographical frame qualified by many scholars as the most avant-garde. It contains also interesting anecdotic and useful historical and philosophical data, which facilitate its reading through technical material. Although it is not the intention of the book to propose a new historiographical method, it indeed is revolutionary as it breaks with the traditional points of view.

The technical references are explained and are generally taken out of the pure mathematical context, although there are always appendices developing the pure mathematical face and references to my work and publications on the purely mathematical part.

The main character in the first part of the book is P.M.S. Blackett, Nobel Prize awarded and political and academic figure in Britain. Several studies have discussed two famous reports Blackett wrote during the Second World War, which constituted the base for all applied mathematics to strategic war, and moreover the interaction of scientists in non-scientific.

However, I analyze also the relationship between Blackett's formative studies on cosmic radiation and the methods he developed for discussing his application to war, that led to the emergence of the discipline that quantified management decisions and planning, which have not attracted the attention it deserves. The same could be said about the epistemological aspects on Blackett's work, and on the intellectual (rather than personal or anecdotic) history of his position vis-à-vis the emergence of Management Science as a discipline.

This work relies on the basis of substantial archival work in the military files. I have tried to discuss these areas and also explore the epistemological limits of a paratextual activity conducted by Blackett with respect to his involvement in the creation of Operational Research, in order to get an effective and optimal influence in the executive levels.

For the second part of the book the study is centered in the main figure of R. A. Fisher, who has also attracted the attention of some scholars; however, and understandably, to a much lesser degree than Blackett. Studies on him have concentrated on biography or internal (and important) aspects of the history of statistics; in particular, giving an emphasis to his contributions to genetics and the theory of errors.

A recent interesting study of Segal [2003] has deepened the subject introducing considerations on the notion of information in connection with Fisher's work. This work is part of a group of pioneer works, which have made an effort to introduce new levels of analysis in the philosophy of mathematics. In this work, using Fisher's texts as well as his correspondence and personal papers (most of which still wait to be properly organized and catalogued), I attempt to analyze the epistemic steps followed by him in the development of his original ideas on the statistical treatment of experimental data, and on his notion of experience and experiment. In addition I also make an analysis of the process of designing experiments using semiotic tools, which are the key to develop strategies and plans in complex organizations nowadays.

The third author considered in this work is L. V. Kantorovich, a Nobel Prize winner, as Blackett, but in his case for the implications of his work on economics. In addition, Kantorovich was a very distinguished pure mathematician with foundational work on Functional Analysis. I pay particular attention to Kantorovich's work on linear programming regarding it as a discursive enterprise. For this reason, I consider in some detail the achievements of Kantorovich in the field of linear programming, analyzing his narrative with linguistic and epistemic tools.

Thus, this work considers the philosophical and mathematical issues of these three important cases in management science and problem analysis and solving: The first part concerns the emergence of Operational Research, this is included in chapters 1, 2 and 3, subject that is closely connected to strategic innovation as seen in the historiographical frame. The second part, treats the contributions of R. A. Fisher in the field of Statistics and the Design of Experiments as a basic generation of information for tactical decisions; these are discussed in Chapters 4, 5 and 6. In the third part of the book I discuss the work of L. V. Kantorovich in the field of linear programming as a fundamental tool in mathematical modeling to quantify executive decisions; this is the topic of Chapter 7. Finally chapter 8 is the extrapolation of these theories in a Latin American case and the analysis of the influence of these views in this international environment.

London
November 2005

Chapter 1
An epistemic approach to the emergence of Operational Research as a business tool.

Overview of Operational Research and its emergence in Britain.

> Operational Research Sections have been established at Headquarterss of the Bomber, Fighter and Coastal Command. The purpose of these sections is to provide Commanders in chief with trained scientific staff who can collect data and undertake research into an analysis of technical, tactical and general operational factors from which can be drawn deductions and lessons which will serve to guide C-in-C and their staffs in the conduct of future operations and in formulating the operational requirements. [Public Record Office, AIR2/5352]

A main feature in business development is communication. However, analytical methods and technical resources are also vital. In this sense, in the case of the emergence of Operational Research as the mathematical discipline of optimizing corporate issues, the balance between the utilization of mathematical tools and its effective communication was the key of success, not only to accomplish successfully projects in the military organization (as executives) but also in the emergence of the discipline as such.

Therefore, in this chapter I will explore three crucial issues in the emergence of Operational Research in Britain: a) the nature of Operational Research and the involvement of Blackett as the main protagonist in the mergence of the discipline acting as a consultant, b) the links of Operational Research with cosmic radiation techniques and other physical methods, and c) the historical-textual approach of Operational Research during the emergence of the discipline in Britain. Several books and papers have been written about the emergence of Operational Research, however the ones concerning Blackett's work in the emergence of Operational Research during the Second World War years are in much smaller numbers, for example one of them,

[Rosenhead, 1989] on Operational Research, only refers to the period after the Second World War, [Rider, 1992] links Operational Research with other areas, such as game theory, or [Brentjes, 1994] with linear optimisation. Operational Research in geographical areas outside the U.K. has been considered by [McArthur, 1990], and [McCloskey, 1987] and [1987a] for the United States[1] or [Kaijser, 2000] for Sweden. Although this last author discussed Blackett tangentially he regards him as the main character in the emergence of Operational Research.

Nevertheless, whether Blackett was or was not the initiator or main precursor of the basic approaches of Operational Research is a topic still debated by historians of science, (see [Rau, 2001] or [Kirby, 2003]. For some, his prominence has a strong political element and they propose it dominated his scientific work, suggesting, to some extent, that he did not contribute meaningfully to Operational Research. For others, Blackett is the major figure and the genuine 'father' of the discipline. This question is debated sometimes in a rather partial way, arguing whether the term Operational Research was mentioned in the same sense before him, whether the activities done by Blackett with the military were performed previously by others, or whether the techniques used by him constitute innovations to the discipline.

In this chapter, I consider three different points of view: firstly, the institutional, secondly, the technical including the mathematical and finally the textual activity. These perspectives will be considered in the context of the social environment that surrounded the early years of Operational Research in Blackett's work.

When trying to look for the beginning of Operational Research one can trace its roots far back in time. Asking for an *absolute* beginning of Operational Research is not entirely meaningful as one can often find a pre-history or precedents to events that can be taken as the "real" beginning of the discipline. For example, we can trace from

1 In the United States, the latter points to Thomas A. Edison and his work as head of the U.S. Naval Consulting Board as one of the precursors of Operational Research.

Archimedes to Galileo scientists working in military affairs and enlarge the list considerably after Galileo, thus reducing the problem to the officialisation of the term or the professionalisation of the discipline. One commonly accepted answer has been to take the professionalisation of the discipline, the foundation of the society of Operational Research or the date of the first issue of the Journal of Operational Research the starting date. Blackett recognises that:

> This group [the Operational Research group] was not the first group of civilian scientists studying operations, but it was certainly one of the first groups to be given both the facilities for the study of a wide range of Operational problems, the freedom to seek out these problems on their own initiative, and sufficiently close personal contact with the service Operational staffs to enable them to do this [Operational Research, Recollections of Problems Studied, 1940–45, reprinted from "Brassey's Annual", 1953, pp. 88–106, Blackett Papers, Royal Society Archives, D59].

Therefore, in order to formulate an alternative option for answering that question I will choose a different perspective, such as the proposal of an epistemological approach through which the events may be better discussed. In this paper, using a specific cognitive framework base on constructivist theories, I intend to stress the historical perspective of P.M.S. Blackett's work on Operational Research.

The nature of Operational Research: business development.

From its beginning Operational Research has had an engineering rather than a mathematical flavour, for example there are in non-mathematical journals of 384 articles in reference with Operational Research since 1950. The journals exclude those of Operational Research and management science (management science is, according to Dantzig [1963], another name for Operational Research), if we include these last journals the number of articles jumps to more than 5000. A list of the journals reviewed can be found in the appendix 1.

[Rau, 2001] agrees in this view, stating that Blackett developed the discipline into an activity more proper of the area of engineering. [Mindell, 2000] has given a wide account on the question, including activities of radar defence in connection with Operational Research. Behind these developments lies the idea of introducing more firmly concepts of mathematics, or I would say, a formal mathematical approach, in the applied activities considered by engineering. This is definitely not new; it happened, for example, with the construction of bridges in Britain in the decade of 1830[2] and in Germany, years before the Second World War, with Felix Klein who, according to [Tobies, 1989], realised the new demands that industry was making on mathematics. Industrial application is not the only case pointing to demands made by praxis to the exact sciences in these years. In fact, in the case of Britain since around 1938, scientists from the Cavendish laboratory at Cambridge were sent to Bawdsey to be instructed on radar procedures (see [Kirby, 1997, 561]).

The former separation of pure and applied sciences, which has its modern roots in Kant, considered sciences according to their theoretical-empirical character and had a very considerable influence in 19th and 20th century (see [Schubring, 1989] and [1989a]). Nevertheless the mathematisation of social sciences in Germany, through courses in the *Technische Hochschule* and the *Technische Universität* in the beginning of the 20th century, (due to the educational reforms achieved by Klein), aimed at areas like finance, mathematical economy and actuarial theory, further retaken in the decade of 1960 (see [Knobloch, 1989, 260–264])

So historically the duality between the theoretical and practical activities was crucial in the development of what we know now as Operational Research. One of the reasons why Blackett was so

2 For example Isambard Kingdom Brunel (1806–1859), was involved in the construction of railway bridges; his last, and greatest, was to be the Royal Albert bridge, crossing the river Tamar at Saltash near Plymouth. The bridge has two spans of 139m/455ft and a central pier built on the rock, 24m/80ft above the high water mark. Brunel also made outstanding contributions to marine engineering with his three ships, the Great Western (1837), Great Britain (1843), and Great Eastern (1858), each the largest in the world at date of launching.

appreciated in the military circles was his capacity of being effective in the theoretical and practical areas.

In 1939 just before the outbreak of WWII Blackett was asked to head a group of scientists at the anti-aircraft command working on a new air defence system based on radar technology [Lovell, 1976], [Kaijser, 2000, 385]. In 1940 A.V. Hill introduced him to General Pile[3] becoming his Scientific Adviser at the H.W. of A.A. Command at Stanmore until March 1941, for only seven months.

In fact, General Pile regarded Blackett as the ideal man for the job, he wrote: “In Professor Blackett I had got hold of a first-class scientist, who was going around the country experimenting in the company of the best soldier-gunnery expert I could find” [Pile, 1945, 160, quoted from Lovell, 1976, 56], also in [Fortun, 1993, 602–603]. So Blackett returned to work closely with the military, and he devoted considerable attention to developing and institutionalising a relationship of trust with his military sponsors [Rau, 2000, 218]. Again, according to Rau, Operational Research by 1940–41 transcended its engineering origins and was transformed into a “scientific” activity, I would go even further, suggesting that, it was formalized and circulated in documents, which reached a considerable large audience. This step, as we shall see, was of some importance.

In this same way we can find how the scientific (quantitative) analysis permeated in industrial and managerial activities after this pioneering work. Nowadays, in Latin America we can spot special cases of consultancy that enhance the values of reports based on this kind of analysis, for example PEMEX in Mexico, where consultancy firms like McKinsey are often hired to evaluate projects, but moreover, “All international business activity involves communication" (quoted from Martin and Chaney, 1992, p. 268).

3 Sir Frederick Alfred Pile, was born in 1884 became captain in the army in 1914, major in 1916, Lt-Colonel in 1927, colonel in 1928, Major General in 1937, Lieut-general in 1939 and General in 1941. He was the commander in the A.A. Command between 1939–45. These documents can be found in the Lidell-Hart Papers in Kings College London.

War, Politics and Mathematics

During the war, there was the necessity on both sides of accounting their losses. This was a common practice in the military activity, for example, as a way of rendering arithmetical artifact, it was stated that in Hamburg alone 50,000 people were killed up to 1943 (see [Pearce, 1992, 332]). So in its early years, Operational Research was built using a similar approach, i.e. a quantitative one. If one takes into account this characteristic, one can explain that the mathematical sophistication or specialization was not an obligation for a group more interested in quantify and identify variables. In fact, one of the main features in which several authors coincide is that the discipline took their collaborators out of several academic backgrounds, at least of those of the group that "founded it" [Forder,2000, 2].

To complement this view of the early years of Operational Research, on the American side, since 1934, the National Research Committee, employed people like John Von Neumann and Richard Courant to develop statistical methods, numerical analysis and computational techniques later used officially for Operational Research (see [Owens, 1989, 287]) with the use of these methods one can see the discipline with a 'quantitative' approach. Moreover, Operational Research can be seen as "Operationalistic" in the sense that its scientific concepts are defined in terms of the "operations" used to identify or measure them. For example the use of probability as an "operation" was used to identify the proper activities of Operational Research, but the use of probability also became a scientific "concept" of Operational Research. In other words, the theory is the praxis as well. That is why our exploration of the epistemological limits of the discipline are supported by a clear definition of scientific and technological activities *en tant que* knowledge.

Quantifying was of the utmost importance in order to take decisions on war matters, but more important was the interpretation and manipulation of those figures in order to back the decisions taken; so Operational Research grew quickly as a political weapon which assisted the decisions taken by "gusts of emotions" and in 1941 was recognized as a vital activity by the Air Ministry, [Kirby, 1997, 565]. As a matter of fact Operational Research sections were institutionalized to the point of being part of the bomber, fighter and coastal command of the Royal Air Force [Air Ministry, 1963].

In the view of [Kittel, 1947, 151] Operational Research's main activity has to support executive decisions on the basis of quantitative methods. Nonetheless, Blackett advances come through a combined effort done by both sides, executive and scientific, working together [Blackett, 1962]. In fact, even if many decisions were taken by "gusts of emotions", as the same Blackett said, the validation and legitimization of that decision was given by a rational and 'scientific' method, which was led by Operational Research.

A Definition of Operational Research.

Even in the very recent books in the history of Operational Research, the definition of Operational Research is frequently revisited. However, a precise historical discussion of this matter is, by norm, left hanging often unqualified. Several Researchers working in this area have developed notions, sometimes interesting notions, of what Operational Research is, but never went into the historical details on the definitions of the discipline according to the main protagonists.

For instance, in [Kirby, 2003, 3] the definition of Operational Research is taken from the Journal of Operational Research, however Kirby quotes a definition of Operational Research produced some five years later than a first definition of Operational Research was given in the same journal, this first version differs, to the latter as this one includes some modifications. This first definition that appeared in the Journal is the following:

> Operational Research is the application of the methods of science to complex problems arising in the direction and management of large system of men, machines, materials and money in industry, business, government, and defence. The distinctive approach is to develop a scientific model of the system, incorporating measurements of factors such as chance, risk, with which to predict and compare the outcomes of alternative decisions, strategies or controls. The purpose is to help management determine its policy and actions scientifically. [Operational Research Quarterly, 1962, 13 (3): 282]

It can be argued that some of the topics listed in the definition had been considered by people working in their respective fields, by scientists and by other thinkers long before Blackett. However, it must be noted that Blackett's work was focused on the managerial and political sides generating knowledge in this sense; at the same time (Blacketss' time) there were other groups in the Air Ministry but none of them could generate the "grand narrative" mentioned by Lyotard [1988], that generated meaningful knowledge in which other researchers stood to develop the area. However, it seems advisable to discuss other notions of Operational Research given at that time with more precision.

As a primarily practical tool, Operational Research was defined by its objectives and functions. The general tone of the objectives of Operational Research did not change until the end of the war (1945), as we can see in the report "ORC functions" (Operational Research Committee) dated 3rd January 1946. It states the Operational Research Committee's functions as: "To facilitate the fullest liaison between all concerned with the Operational Research regarding air warfare and with associated technical and intelligence activity" [Public Record Office, AIR 20/1728]. Although this report is from 1946, the functions and objectives were quoted and retaken from documents from the early years of the 1940s decade:

A summary of the functions of Operational Research can also be extracted from the reports issued by the respective sections of the Bomber, Fighter, and Coastal Command of the Royal Air Force. The duties are as follows:

> Bomber command (from notes on duties of Operational Research Officer from 13 September 1941). [4]
>
> > Analysis of interceptions and losses to show effect of height concentration routering and moonlight.
> >
> > Analysis of bombing results from night photographs to show percentage of aircraft reaching the target under varying conditions, and deductions there from.

4 Italics are mine·

Study of night bombing operations in order to deduce the existence of dummies and pin-point then on the map

Analysis of day operations with particular reference to their effect in bringing about a redistribution of German defences

A study of the German defence system with a view to discovering weak points which can be exploited

A study of the Operational returns rendered by units in order to ensure that all the information therein is of value and is presented in such a form as to provide good statistical material

Analysis of casualties to assign the causes e.g. enemy fighters, flack, engine failures, etc.

Fighter command (Syllabus of work for ORS FC HL/CMG from 8 October 1941).

Cooperate with intelligence and advise on the obtaining analysis of data relating to night operations in order to determine:

Reasons for success or failure of our defensive operations related to enemy effort and tactics weather, conditions of visibility, and the area concerned.

The most profitable employment of the various defensive weapons or combination of weapons at our disposal (e.g. A.I. or cats eye fighters on "free-lance patrol" or under various forms of close patrol, turbine-lite interception aided by SLC searchlight; mutton; albino; fighter nights) according to the conditions prevailing during the attack and to the tactics employed by the enemy

The degree of efficiency achieved or weakness of any specific method of attempting the destruction of enemy aircraft (e.g. the successive stages of a GCI/AI interception are affected in varying degrees by several factors).

The success or otherwise of night operations such as "intruder".

Study the general problem of recognition of friend from foe, to cover both, the direct methods such as markings on aircrafts, resin lights, etc. and indirect methods such as accurate plotting with retention of identification. Investigate the possibility of routering bombers in such a manner as to ease the general problem of recognition by night.

Advise and cooperate with CSO or other branches on the general problem of taking off and landing aircraft under conditions of bad visibility.

Conduct ad hoc investigations as necessary into short term or immediate queries which arise.

Coastal command (from the Programme of ORS CC 15 September 1941)

Analysis of air attack on u-boats

Analysis of sighting of U-boats

Performance of ASV

Analysis of methods of hunting U-boats

Analysis of long range enemy aircraft activity

Analysis of effectiveness of A/S air effort

Study of camouflage of A/S aircraft

Study of performance of radio navigational methods. [PUBLIC RECORD OFFICE, AIR20/1728]

As we can see, typical words in the field of modern Operational Research like 'optimization', 'maximise' or 'minimise' are not the main objectives of the list. The idea at that time was to create a causal link using a low number of variables. The main terms with which these groups worked were: 'study', 'analysis', 'determine reasons' and 'cooperate' which, in turn, have very different meanings. But this is precisely the kind of approach that we use in modern business planning, especially in cases like the Latin American one.

Blackett's personal view on Operational Research can be found in one of the working papers elaborated for future publication notes:

> The objective of operations analysis is to identify the significant variables in the operation of any system, evaluate their significance with the aid of observational data supplemented by theory, and draw conclusions as to what variations will lead most effectively to the desired result [Blackett Papers, Royal Society Archives, D113].

The above quotation implies the task of Operational Research as that of finding the main variables. It is interesting to note that the activities of these sections were focused on the technical part of the operations. As said before, in the beginning, Operational Research had a technological objective more than a scientific one. Blackett's activities could best be described as "technological", rather than a "speculative" or "scientific". This is reflected in his writing, as we will see it in the following sections. The activities began with the defence of Britain by radar systems and, later, when dealing with the mechanical devices for sightseeing, for example, Mark XIV[5]. In this sense, and from a historiographical-textual point of view Operational Research is a technique to deal with objects and explain situations of 'reality', and not primarily with the development of theories. Technique in this case achieves the knowable objects first than science, that is why in terms of interaction with reality Operational Research can be seen as a technique and not as a scientific development. Therefore the knowable objects and the cognitive subject cannot be specified unless they are taken as pre-epistemic entities. In other words, to analyse properly this case we exclude the cognitive act from the technique. That is why precisely, in the case of applied mathematics, epistemic frames such as the one developed by Bloor [1976] (called Edinburgh School) leaves a lot of holes in the explanation of Operational Research as an epistemic discipline, concerning business development.

In this sense, Operational Research takes also a different tone, the one of technology, because it is also concerned with the explanation of the activities of war operations as situations of reality (the same happens in business), it is important to note that Operational Research is not concerned with the development of military theories or finding the solution to some numerical equations. Blackett was involved in such actions as pointed out in the objectives of the Committee of Scientific Advisory to the Military Intelligence": the so-called "Tizard Committee"[6].

The Committee had a well-defined and specific target: "To consider how far recent advances in scientific and technical knowledge can be used to strengthen the present methods of defence against hostile aircraft" [Public Record Office, CAB 86/3].

5 Depth charge as a submarine torpedo to counter arrest the U-boats activites.

6 Which official name was "Committee for the Scientific Survey of Air Defence". the first meeting of the committee was held in 28th of January 1935. Present were Sir. Henry Tizard, Blackett, A.V. Hill, Wimperis and A.P. Rowe. See archives of Public Record Office and Blackett papers in the Royal Society Archives.

It is in under this view that it comes clear that the generation of knowledge in the applied mathematics area is given in the writing activity and not in the actions taken before or after any theoretical consideration. In fact, theoretical considerations are not part of the narrative of Operational Research. It is like this how we can explain why the purpose of the Operational Research pioneers was to generate a writing activity; a literary product that based on natural scientific theory could be inserted in the judicial and institutional planes, in which the narrative identity looked for the permanence of an argument more than a decision taking.

In this sense, Blackett is creating a reality principle that transcends the borders of military activities, but not the borders of scientific activities. Within this frame it is clear also why other scientists or precursors of Operational Research did not follow techniques developed by Blackett. The 'Great Narrative' or the essence of knowledge in Operational Research (and other applied sciences like Statistics) is aimed, directed and instructed, as Gerard Genette [1972] states, to the identification of a narrative identity, which is of crucial importance to determine the epistemological features of the discipline.

The narrative gives shape and body to a cognitive frame in which methodologies are developed. These methodologies are without doubt closely connected to the philosophical approach of the matter, and this is why some authors confuse the foundations of a discipline with their methods, (see [Hand, 2001] where he attempts to discuss on the foundations of 'Data Mining' and ends up describing disappointingly the methods and algorithms of that discipline).

In the case of Operational Research the methodologies are discussed apart from the epistemological ground and these methods can be followed in the mathematical and physical areas, that is why a philosophical frame as the empiricist one, leaves a lot of activities of Operational Research without proper explanation. In fact a constructivist theory explains in a clearer way the relationship between the methodological and epistemological activities in the applied disciplines.

The geometric nature of Operational Research: Back to Cosmic Radiation.

The nature of Operational Research according to Blackett is geometrical. He attempts, at all times, to build linear approximations and tries to compute differential coefficients in order to explain operations of war. In order to do this he proposes methods like the a-priori and the variational [Blackett, 1948] and [Rider, 1994, 838]. Although not emphasized by Rider there is also an interesting mixed method as indicated in section 5.3:

A desire derivative $\frac{dY}{dX_1}$ may often be obtained from another Operational derivative $\left(\frac{dY}{dX_2}\right)_{obs}$ by using (a) a theoretical or (b) an experimental relationship between the assumed causally related increments dX_1 and dX_2 [Blackett, 1948].

Formally one can express the above method by the relation

$$\frac{dY}{dX_1} = \left(\frac{dY}{dX_2}\right) \times \frac{\mathrm{d}X_2}{\mathrm{d}X_1}$$

where $\frac{\mathrm{d}X_2}{\mathrm{d}X_1}$ is determined either theoretically or by special experiment [Blackett 1948].

And as [Blackett, 1948] puts it:

Expressed in geometrical language, the first task of Operational Research faced with a new operation is to investigate the shape of a multi-dimensional surface

$$Y = F(x_1 \ldots x_2)$$

surrounding the point corresponding to a past operation, and to use this knowledge to predict the properties at a neighbouring point corresponding to a future operation [Blackett, 1962, 32].

With the tools identified above, he can infer the need of linearization of the models explaining the operations of war. Here we can find a similarity with his work on cosmic radiation and, moreover, a connection between this methods and the classical tool of Operational Research: linear programming.

This idea seemed to have escaped to authors discussing the history of Operational Research; many of them have traced trace Operational Research ideas to the 1770s (as discussed by [Grattan-Guinness, 1994, 43]) in connection with the work of Fourier and to the early days of the theory of convex spaces[7], or the use of probability in applied situations can be traced far back to Euler and his theory of errors and to the origins of the so-called mathematics of the late 18th century (see [Sheynin, 1971, 45]),but it was never mentioned the connections with Blackett's same work on Cosmic Radiation.

The work and views of Blackett in the early years of Operational Research (1934) was concerned with linearization and extrapolation of methods normally applied in experimental physics, a field in which he was immersed since the beginning of his research activities. He participated around that same year of 1934, in the Committee for the Scientific Survey of Air Defence, with Sir Henry Tizard, and from 1935 with Robert Watson-Watt in his work on the radar system; in 1936 in the establishment of the Research station at Bawdsey; in 1937 with the radar experimental station at Bawdsey Research station; in 1938 together with A. P Rowe and E.C. Williams, analysing more deeply the radar operations and, finally, in 1939 in the activities of the Operational Research section in Stanmore with the fighter command of the Royal Air force, in all these years the extrapolation of physical techniques is present in his work, (for more on Blackett's activities see [Larnder, 1978]).

7 The work can be found in the Oeuvres de Fourier vol 2 of the Academie de Sciences, Paris.

From Cosmic Rays to Operational Research and planning.

> We can infer that some of the ideas in the field of Operational Research permeated from Blackett's previous studies on cosmic radiation. An important source for extrapolation and comparisons can be found in his cosmic rays book [Blackett P.M.S., 1935][8].

Empirical methods of Operational Research are linked to physical methods used by these scientists before the war but in his last case there are also mathematical tools, connected with his cosmic radiation studies. For example, vector calculus[9], differential coefficients and Poisson probability distribution[10]. The exponential function, which is in a way the continuous case of a Poisson distribution, is used to calculate sightseeings radius in the Operational Research field. This is clearly visible in some of Blackett's personal notes, for example in the one concerning the use of statistical methods [Blackett Papers, Royal Society Archives, B24] where he uses the exponential method, the concept of curvature, and what he calls the check by "used" x' y' coordinates on the curved proton tracks; similar remarks can be found in other related working-papers written [Blackett papers, Royal Society Archives, B22] in connection with drafts for a thesis prepared for a fellowship at King's College.

8 La radiation Cosmique, Fondation Singer-Polignac, Hermann & Cie Editeurs, Paris, (1935).

9 Using directional derivatives and building tangent planes in certain points of the 'surface' generated by the model.

10 In general, the application of probabilities.

A First Connection with cosmic rays: Equilibrium Theorem.

In his studies on cosmic radiation, Blackett depicted, as seen in the following illustration (next page) taken from [Blackett, 1935], a visualisation of trajectories in a dynamical system. At the end of the day, in a dynamical system probably the most important question is where is it going to end i.e. its asymptotical behaviour, that is If it will collapse or diverge, or reach an equilibrium. In Operational Research there are also trajectories and similar parameters to analyse, as seen in the following photographs, one used by the Operational Research sections as an information input for analysis and the other in Cosmic Radiation Studies.

In Operational Research there is also an attempt to look at operations of war as dynamical systems. Since the 1920's these ideas were actively pursued by mathematicians. Birkhoff formalised results in this area in his famous book of 1927 [Birkhoff, 1927] where he analyses the asymptotic behaviour when the process (a dynamical system) is iterated. A clear example of this is what Blackett stated as "an equilibrium theorem". Clearly I am not suggesting that a physicist like Blackett was then conversational with Birkhoff's work, but the ideas Birkhoff put into his book were widely discussed by physicists and are directly related to cosmic radiation analysis.

In this theorem Blackett implies that a system iterates, as he says, "in the course of repeated operations" and he assumes that the number of iterations is large and concludes that the system is stable as "the result will be a gradual approach to a tactical state with certain maximum properties". The theorem is contained in the section of [Blackett, 1948] entitled "An Equilibrium Theorem":

> It is clear that an intelligently controlled operation of war, if repeated often enough with reasonable tactical latitude allowed to the participants, will tend to a state where the yield of the operation is the maximum, or the negative yields (losses) a minimum. Exceptions will of course, occur for various reasons. For instance, the tactics of the operation are restricted by too rigid orders, or if the yield of the operation is not known by the participants.
>
> This result clearly follows from the fact that in the course of repeated operations by many different participants, most of the possible variations of tactics will be effectively explored, so that any large derivatives will eventually be discovered,

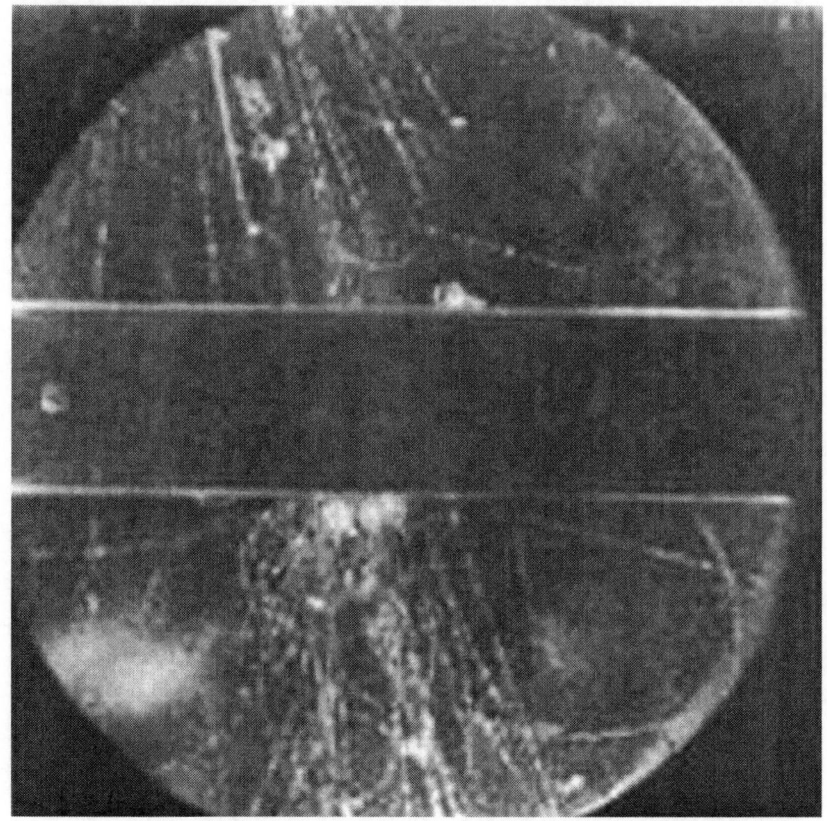

HAMBURG
K4165

> and given intelligent control, improved tactics will become generally adopted. The result will be a gradual approach to a tactical state with certain maximum properties, that is, one in which what may be called the free tactical derivatives are nearly zero [Blackett, 1948, 33].

And not only the vision of a war operation as a dynamical system, but the proposition, just as in his cosmic rays studies, of a multivariate function that was seen as a multidimensional surface in which parameters were measured to figure out the nature of the problems.

Second Connection: The Problems.

As treated by many authors on the history of Operational Research, like [Rau, 2001], the problems in the Operational Research Section began with the radar data predictor and the night defence of London concerning the Anti Aircraft (A.A.) Batteries to which radar sets had been given. The Operational Research Group attempted to weigh up the advantages and disadvantages of the, in that time, current deployment of the available hundred and twenty or so guns in the thirty four-gun batteries.

Another problem was to calculate the effectiveness of the A.A. gunfire, but the most famous problems were those concerning the U-boats attacks in the Atlantic. In none of the histories of Operational Research consulted, there is an analysis of one of the most important and illustrative problem: the "convoy size", although mentioned by many, like [Lovell, 1976], it has never gone beyond quoting it from the original [Blackett, 1948]. The analysis performed here has its core, not in the technical part, but in the epistemological one.

In a paper by Blackett of February 5th 1943, entitled "Progress of analysis of the value of escort vessels and aircraft in the anti u-boat campaign", he refers to the analysis done on the optimal convoy size problem:

> An examination has been made of the attacks of 77 convoys during the period 1941-42 in which the approximate number of submarines in the u-boat pack was known. As expected, the statistics show that the number of ships torpedoed per

submarine present in the pack decreases markedly as strength of the naval escort is increased. The figures show that an increase of the average escort strength from 6 to 9, would have been expected to have decreased the losses by about 25%.

During the last 6 months of 1942 the losses in the N. Atlantic convoys were at the rate of 210 ships per year. The number of escort vessels operating was about 100. The analysis quoted in the last paragraph shows that, had there been an additional 50 escorts available the losses should have been reduced by 25%,i.e. 52 ships.

During the last half of 1942 the transatlantic escorts probably sank 8 u-boats and damaged 8. Taking the latter as equivalent to two more sunk, we see that 100 escorts sank u-boats at the rate of 20 per year.

The escorts saved the shipping which would have been sunk by the u-boat if this would not have been sunk first by the escort. The shipping casualty figures in all areas show that about 0.4 ships were sunk per month by each Operational U-boat in a 9 month period this would amount a total of 3.6 ships.

Combining the above figures we find that 100 escorts saved 20(3.6)=72 ships [Blackett does not say it but it is in a 9-month period] and continues saying that this gives a rate of 0.7 ships saved per escort vessel per year.

From the result of the last two sections one sees that the total saving of shippings in the last half of 1942 by each escort vessel was probably at the rate of nearly 2 ships per year.

And later on, in his paper:

From August to December 1942 the data covered 80 days of shadowing and 21 convoys. From these tables it is found that there were 43 days reckoned dawn to dawn, when shadowing by a pack took place while no air cover was available, and during these days 75 ships were torpedoes by packs of average size 5.3 u-boats.

There were 38 days of shadowing in which air cover was available amounting in all to 147 sorties, in which 24 ships were torpedoed by packs of very nearly the same average size.

Thus in the absence of air cover one would have expected 75 (38)/43=67 ships to be torpedoed. The saving due to the 147 air sorties was therefore 67-24=43 ships or 1/3 of a ship per sortie [PUBLIC RECORD OFFICE, CAB 86/3 AU (43) 40]

In the first paragraph of this long quotation an important thing to be noted is that he expected some concrete results; the same happens in cosmic radiation studies. One has a dynamical system that models reality via: 1) a configuration space (a manifold in the mathematical sense), 2) a vector field (which in this case can be translated as the law of change) and 3) a metric *dx, dxdy* or *dxdydz* (in the case of a volume). From that model, one has expectations that are to be confirmed later by experimental evidence, i.e. by using statistical tools.

In the above paragraph Blackett explicitly showed that the physical methods he had used in cosmic radiation studies could be extrapolated to Operational Research. In cosmic radiation statistical methods began to be of the utmost importance due to the construction adopted for a description of the nature of the phenomena. In the same way knowledge is constructed in Operational Research through this global analytical step; that is, abandoning the notion of reporting individual cases.

Just as in cosmic radiation, Operational Research Sections tended to analyze the observations in a quantitative way, but in order to do so, the information had to be addressed to the Operational Research Section in a statistical format. Therefore, they had tables of data, for example of hours of flight, number of torpedoes launched, number of u-boats sunk, etc. and the way of presenting reports, as seen in the example above, shows the statistical treatment and the experimental methods used by Blackett and the parallelism with cosmic rays. It is important to note that Blackett is not concerned with the statistical methods, but with physical methods followed by a mathematical model, in which statistics play a valuable though auxiliary role.

Another important similarity is the use of an experimental-observational method to draw conclusions, exactly as in studies of radioactive decay performed by Blackett in some previous years, as he finishes the report by correlating the results with an observation that came out from the experiment: "The saving due to the 147 air sorties was therefore 67-24=43 ships or 1/3 of a ship per sortie".

Now epistemic part comes closely related to the narrative style, in which one of the most important parts is the legitimisation. This is achieved in the theory itself, which comes as a tradition of modernity; Science itself is a body not well defined or known, but trustworthy. He never mentions authors or any other historical fact; he only refers to numbers as a token of truth.

In fact, the Heideggerian discussion [Heidegger, 1995],of the visible and its relation with truth makes sense in the process of legitimating the argument. Visible in this case is not visible at all, but it is also referred to observations, in fact observation taken out of experimental situations, or what Blackett calls "Operational Experiments".

Third Connection: The Probability Models.

Operational Research activities could be modelled statistically in a classical way by a normal, uniform or binomial distribution[11], but the main statistical tool mentioned in these reports is the Poisson distribution.

The Poisson probability distribution was widely used in particle physics before the war; in fact, there is a connection between Poisson[12] distribution and cosmic radiation. This distribution, especially used by Rutherford[13] and Geiger, can also be deducted directly from the experimental data on cosmic rays and radioactive decay models for a short term period; based on differential equations, and not only as the limit of binomial distribution [Rutherford, 1930].

In the radioactive decay phenomena, one classically uses a differential equation to describe it. This works perfectly for the long term phenomenon: it can forecast the amount of disintegration at any time, t, and the half life of a radioactive mass. However, for very short intervals, the differential equation does not work sufficiently well. In this case, the Poisson distribution pictures the event more accurately.

11 As in the work of Pearson and Fisher

12 One might recall from statistics theory that the Poisson Probability distribution formula is

$$P(x) = \frac{l'e^{-l}}{x!}$$

With parameter λ.

13 Blackett worked with Rutherford in the previous years in Cavendish Laboratory.

The radioactive disintegration problem can be modelled in two different ways: a) assuming a deterministic model, which will adjust to the long-term period, and b) assuming randomness and this will adjust to the short-term period.

In the deterministic model, we discuss the problem in the classical way, which is considering some amount of Radium which at some time t=0 has a mass M, beginning with

$$\frac{dM(t)}{dt} = -kM(t)$$

And finishing with,

$$M(t) = Me^{-kt}$$

This differential equation allows us to forecast the mass at any time t.

Assuming randomness, we are interested in the same physical experiment, but in a very short time, for example, in less than a second. The times at which these instantaneous decreases of mass occur are not deterministic. However, if one duplicates this experiment many times, one would find that the relative frequency with which 0,1,2,... alpha particles are emitted during a specified period remain "relatively constant". At this point, the problem was not to find the exact number of particles which are emitted after t seconds, but to find the next best thing, the probability $P_n(t)$ that n particles are emitted after a time interval of length t.

To construct the model we first have to make some assumptions:

i. If $0<t_1<t_2$, the number of alpha particles emitted during the time interval $[t_1,t_2]$ does not depend on the number of particles emitted during the time interval $[0,t_1]$. It is not an historical event.

ii. The probability that exactly one alpha particle is emitted during the time interval $[t,t+h]$, for $h>0$, is $\lambda h+o(h)$, where λ is an unknown positive constant, and $o(h)$ has the property that

$$\lim_{h\to 0}\frac{o(h)}{h} = 0$$

iii. The probability that more than one alpha particle is emitted during the time interval $[t,t+h]$, is $o(h)$.
iv. For every n, $P_n(t)$ is differentiable in t, where $P_n(t)$ is the probability that exactly n particles are emitted during the interval $[0,t]$.

And using differential equations, and some algebraic manipulation, we obtain a system of iterative differential equations from which the general solution is derived:

$$P(x) = \frac{1t^n e^{-1t}}{n!}$$

Which, coincidentally, is the Poisson probability distribution.

Now we can see that the Poisson distribution model precisely explains a physical experiment, and could be used to predict, in a practical sense, phenomena performed in a "real life" scenario. Rutherford and Geiger collected the data of a table. They observed the number of alpha particles emitted during time intervals, which were 7.5 seconds of duration. They made 2608 (independent) observations on such intervals i.e. they observed the number of alpha particles emitted during 2608 disjoint intervals. Although the results are not identical, it is impressive that the model (Poisson distribution) gives similar results to the ones afforded by the real physical experiment.[14]

In this example, one can observe that again, the tools to develop this kind of model were differential equations, statistics and probability theory, which makes us conclude that there is in Blackett's approach a deep link between models for the study of cosmic rays and Operational Research schemes.

14 Rutherford with Geiger developed calculations using the Poisson distribution to evaluate the number of alpha particles emitted during an experiment. These are distributed as a Poisson process. Many of the experiments of this kind are modelled and solved with differential equations, and it is important to note that for short term periods, the probabilistic method is a very good approximation to the phenomenon.

In fact, tables of data like those generated in cosmic radiation studies are commonly seen in the reports of the Operational Research sections during the Second World War. In addition, the use of differential coefficients and Poisson distribution as the main tools in the Operational Research sections gives us still more evidence of the parallelism between the physical methods of cosmic radiation experimental Researchers and the scientific advises given to the military during the war. We have not been able to find this connection in the relevant literature on the history of Operational Research.

It should be emphasised that analysis on the methodology of Operational Research had been attempted before. One of the most consistent to be found is in [Eilon, 1975]; the author discusses Operational Research methods using an epistemic frame based on Popper's conception of scientific enquiry. After this section one can see how a theory like the Kuhnian does not apply at all in the case of Operational Research, as it would be difficult to assess a proper limit for "normal science" just to give an example.

P.M.S. Blackett.

I will summarize briefly the biography of P.M.S. Blackett, the person, in connection with his activities linked with Operational Research, especially during the Second World War giving some special attention to points relevant to the arguments discussed in this paper.

Patrick Maynard Stuart Blackett was born in London on 18 November 1897, (in 1942 when he did all the Operational Research work he was 45 years old, leading mostly mid-twenties workers in his team). His father was a stockbroker and his grandfather (Major Charles Maynard, R.A.) was in the military and was in service in the years of the Indian Mutiny. [Lovell, 1976]

As pointed out by [Lovell, 1976] and [Kirby, 2003, 111], Blackett began an early military career in the navy, and when WWI broke out in 1914 he (who was almost 17) was appointed as Midshipman to one of the cruisers, the H.M.S. Carnarvon. He saw action against the German battleships Scharnhorst and Gneisenau in that same year and the following, so by 1940 he already had military experience not only from the academy, but also coming from the battlefield. We could say he was a military man.

In October 1916 he was appointed Sub-lieutenant on the anti-submarine ship H.M.S. P17, Dover Patrol, and then in July 1917 he was transferred to the destroyer H.M.S. Sturgeon. Blackett was promoted to Lieutenant on May 15 1918. He resigned as soon as the war was over.

By that time, the Admiralty decided that 400 young officers whose course in the military service had been truncated at the outbreak of war would be sent to Cambridge for a six-month course of general studies. This established a further peacetime line of contact between military officers and University of Cambridge. Blackett was among these officers and joined Magdalene College in 1919. Thus, his was the case of a naval officer training for science.

In that same year Rutherford arrived at the Cavendish Laboratory from Manchester, Rutherford experimented with Chadwick on fast alpha-particles. Rutherford had a Japanese student, Schimitzu, who had been instructed to use C.T.R. Wilson's cloud chamber technique to get more information on the collision of alpha-particle and nitrogen nucleus. Schimutzu made the operation of the cloud chamber automatic and took a few thousand photographs of alpha-ray tracks [Galison, 1997, 118]; for family reasons he returned to Japan; at the time Blackett arrived to Cavendish. Rutherford asked him to take over Schimitzu's experiments. [Lovell, 1976]

Blackett's first scientific paper on the analysis of alpha-ray photographs was communicated to the Royal Society in 1922 and published in "*Proceedings*", 19 years before his main work in Operational Research.

During the following years until 1924 Blackett worked in this area, where he completed the task given by Rutherford and published his results in "*Proceedings*" of the Royal Society under the title of "The ejection of protons from nitrogen nuclei, photographed by the Wilson method".

Blackett had the opportunity to develop both theoretical and practical insights and combine them into an experimental technique to record and analyse cosmic radiation patterns. As I intend to show, this idea would later be reconsidered in his studies on Operational Research; in fact it will mark a trend in the development of his scientific methodology.

He left Cavendish in 1933 when he went to Birbeck College. In the same year he was elected F.R.S. and became fully focussed on cosmic rays studies. During his stay at Birbeck he published, in Paris, in 1935 "La radiation cosmique" where he gives an account of his experiments on cosmic rays.

In the late 1934, a year after Hitler came into power in Germany, Harry E. Wimperis and A.P Rowe proposed to the Air Ministry to establish a Committee for the scientific survey of air defence. [Rau, 2000, 220].

As pointed out by [Lovell, 1976] (and also by [Kirby, 2003]), in 1938 they created a committee for defence and Sir Henry Tizard[15] who was then Chairman of the Aeronautical Research was appointed to chair this "Committee of Scientific Advisory to the Military Intelligence", the so called "Tizard Committee"[16].

Later Tizard would become an ambassador for studies on Operational Research in the United States, but not before selected staff members of the National Research Defence Committee (NRDC) spent tours of duty at the London Mission [Rau, 2000, 62]. Although, as said

15 Sir Henry Tizard, joined the RGA in 1914 and was transferred to RFC in 1915, between 1918 and 1919, he was Lieut-Colonel and Assistant Controller Experiments and Research in the Royal Air Force. During 1927 to 1929 he was permanent Secretary of the Department of Scientific and Industrial Research and was Rector of Imperial College in the years of 1929 to 1942. He was Chairman of the Aeronautical Research Committee from 1933 to 1943 and member of the Council of Minister of Aircraft Production and of the Air Council from 1941 to 1943. In the years of 1946-1952 he was chairman of the Advisory Council on Scientific Policy and Defence Research Policy Committee. The most important fact in this case is that he was a Development Commissioner of the Air Council from 1934 to 1945.

16 Which official name was "Committee for the Scientific Survey of Air Defence". the first meeting of the committee was held in 28th of January 1935. Present were Sir. Henry Tizard, Blackett, A.V. Hill, Wimperis and A.P. Rowe. These documents are preserved at the Royal Society Archives and at Imperial College London Archives.

by [Mindell, 2000, 29], the "Tizard Mission" in the United States revealed concepts like the "cavity magnetron" to the Microwave Committee, this exchange of scientific knowledge was then of the utmost usefulness for the United States.

In 1937 Blackett moved from Birbeck to Manchester where he remained until 1953 developing the same line of studies (cosmic radiation) but with a better-equipped laboratory than at Birbeck.

In 1939 just before the outbreak of WWII Blackett was asked to head a group of scientists at the anti-aircraft command working on a new air defence system based on radar technology [Lovell, 1976], [Kaijser, 2000, 385]. In 1940 A.V. Hill introduced him to General Pile[17] becoming his Scientific Adviser at the H.W. of A.A. Command at Stanmore until March 1941, for only seven months.

Pile regarded Blackett as the ideal man for the job, he wrote: "In Professor Blackett I had got hold of a first-class scientist, who was going around the country experimenting in the company of the best soldier-gunnery expert I could find" [Pile, 1945, 160, quoted from Lovell, 1976, 56], also in [Fortun, 1993, 602-603]. So Blackett returned to work closely with the military, and he devoted considerable attention to developing and institutionalising a relationship of trust with his military sponsors [Rau, 2000, 218]. Again, according to Rau, Operational Research by 1940-41 transcended its engineering origins and was transformed into a "scientific" activity, I would go even further, suggesting that, it was formalised and circulated in documents which reached a considerable large audience. This step, as we shall see, was of some importance. Blackett was an ineterst man for a post in the military organisation, as he read widely and had strong interests in fields like philosophy, history, anthropology, and psychology [Nye, 2004, 7].

17 Sir Frederick Alfred Pile, was born in 1884 became captain in the army in 1914, major in 1916, Lt-Colonel in 1927, colonel in 1928, Major General in 1937, Lieut-general in 1939 and General in 1941. He was the commander in the A.A. Command between 1939-45. These documents can be found in the Lidell-Hart Papers in Kings College London.

In 1941 there were concerns about the German submarines' threat to merchant ships and also about the difficulties the Coastal Command had to overcome for an effective counter attack. The C-in-C at the Coastal Command, Sir Philip Joubert de la Ferté, took the initiatives leading to Blackett's transfer from the A.A. Command to the Coastal Command in March 1941. It was in these years that he published the first "scientific" paper working in a military environment, in it he explained the activities of an Operational Research Section and, years later, made explicit his objectives: "The first of these (documents) was written in order to inform the Admiralty of some of the developments which had occurred in the Operational Research Sections already established at Fighter, Anti-Aircraft and Coastal Commands" [Blackett, 1948, 26]. This paper was accompanied by a talk to the Admiralty's staff. [Rider, 1994, 838], who commented on this paper, wrote that "Blackett argued for numerical thinking in Operational matters", this argument is directly related to Blackett's concept of science. The general tone of this paper is more informative and aimed mainly to tactical issues, i.e. at improving existing weapons system or production systems, where the methods employed had a scientific and mathematical bias [Kaijser, 2000, 386].

Blackett was in the Coastal Command for nine months. His early talk of December 1941 stimulated the interest of senior members of the Naval Staff. So, in 1942, and until the end of the war, he was transferred to the Admiralty. In January 1942 he became Chief Adviser on Operational Research (C.A.O.R.), and his group was placed directly under the Vice-Chief of Naval Staff. In 1944 the post was changed to Director of Naval Operational Research (D.N.O.R.).

In May 1943 he released a second document, complementing that of 1941, which was an attempt to set out the relevance of these new methods of analysis in the military field for the benefit of new scientific recruits of the Operational Research sections (at this time, there were three sections in the Royal Air Force). This document also had a wide circulation in military circles.

Blackett left the Admiralty in the summer of 1945. Although he was Chairman of the RAF Aircraft Research Committee from 1947 to 1950 and Chairman of the Naval Aircraft Research from 1945 to 1950, his interests shifted from Operational Research to magnetic fields, which was alpha as explained by Nye, [Nye, 1999, 72][18].

Emergence of Operational Research.

It is not my intention to minimise the manifold influences in the emergence of Operational Research. Indeed, the beginning of the discipline is complex and full of visible and hidden roots, such as military, sociological, technological, political, economic and institutional. However, a point of view from which one can understand internal, as well as external factors concerning to business development, in the emergence of Operational Research is an epistemological analysis of Blackett's activities during the war years in Britain.

It is well accepted from the institutional corner that Blackett was a leading founder of Operational Research. As a matter of fact, the first organisation of professionals in the discipline (The Society of Operational Research of Britain) recognised him as such. He was given the honour to write the first paper in the Journal of Operational Research. However, one can give further arguments, besides those given here, which may help to decide whether Blackett deserves -by 'scientific' or other merits- to be regarded as the originator of Operational Research, a point on which there is divergence in the literature, (see [Kirby, 2003]).

To ascertain this, firstly, I will define Operational Research from a historical standpoint. Secondly, I will focus on the explanation of the events related to Blackett's vision of Operational Research in his times, which influenced meaningfully his work in the early 1940s. That is, through the process of generation of knowledge using an epistemological viewpoint.

18 In a thorough analysis Nye discussed the last phase of Blackett's Research work.

An Epistemological Ground.

Since this work is supported by research dispersed over many documents, correspondence, reports, books and articles, it has in this sense elements of a historiographic account. Nonetheless, as a historiographic product a work might induce some self-governing epistemic considerations or simply adjust itself to the changing models of writing and styles regarded as "established" or "accepted" at a given time as a cannon. The self-governing epistemic model, which is the most interesting, must have a structural base in which all the concepts and ideas acquire a meaning. In this work I have used an epistemic construction which is not only the result of a conciliation of many arguments and variables, therefore, it does not refer to an exclusive methodology or point of view, be this social, economic, administrative, institutional military or political, and can be better seen in the convergence of the history of ideas and mathematics.

This work is also the result of an exploration on the relationship between writing activity and subjective experiential knowledge, within the temporal frame of early 20th century science, and taking P.M.S. Blackett and his work in Operational Research as a specific subject.

Therefore one of the points of view used in this work is the analysis of texts generated by a writing activity performed by a scientist, Blackett in the context afore mentioned: knowledge = writing = narrative ⇒ text ⇒ epistemology[19], where by "=" I mean equivalence in the forms and contents of cognitive acts. Epistemology should be understood here as the drawing of planes and a creation of concepts.

19 For 'narrative' I use a concept specially defined by Lyotard in his influential work [Lyotard, 1979], specially sections 9 and 10.

The Texts: a formalization of Operational Research and the communication of results to executive levels.

There might have been some texts mentioning Operational Research activities before 1941, nevertheless it is since Blackett wrote his famous notes during his service in the Ministry of Defence that Operational Research was formalised in its present format. Furthermore, Blackett's Operational Research documents must be credited with the innovation of breaking with the standard format and style of the military and administrative reports hitherto given in the military intelligence organisation.

Sir Henry Tizard recognised this fact fully and in February 21, 1942, when he stated that Blackett's paper explained how to analyse war operations from a scientific viewpoint and "should be circulated in the military organization"[20] [Royal Society Archives, Blackett Papers, D83], Blackett's document became well known to military groups in Britain.

War operations (one of the objects of the discipline at that time), may have been formulated in several alternative forms. However Blackett's writing offered one of the most penetrating and influential, one of the factors involving the generations of these texts could have been the political element that influenced the whole organisation. For instance, scientists wishing to have a position of power in the organization as scientific advisers, or conversely politicians trying to position themselves as "scientist" with a political flag. This idea has been developed by [Kirby, 2003, 314-317] where he considers the impact of Operational Research ideas in the ideology on scientific Research of post-war Labour.

Another circumstance involved in the textual generation is that that the elaboration of models taken by scientists to the political arena as a true novel idea required the backing of personalities of Blackett's stature, in other words, the problems posed by Blackett were of such a nature that it was irrelevant to show methodologies or propose general

20 In the original papers [Blackett Papers, D83-D128], it seems that Blackett only sent a draft to Tizard in that time.

algorithms, in this case it was more meaningful to answer direct questions rather than to offer a discourse on how to "optimise travel time of some convoy" backed on the figure of a well recognised scientist; to round the contextual frame, we can say that the approach of translating scientific concepts was entirely acceptable under the given circumstances (this is part of a classic debate of militaries vs. scientists), this meant that the problems evolved and changed from one stage to another, but at the end (at the time to communicate results), it was out of the question to write something like "The correlation coefficient is 0.87", because it would have lost its essential meaning in the interpretation (i.e. it was out of connotation) and moreover, would have skipped a step in the pragmatics of the discourse, in other words, it would have lost the power of understanding the explanations of reality in the 'real world'.

It is not an easy task to analyse the literary factors involved in the documents, but looking at these factors one confronts a wider range of historical issues to consider, as it is the intention of the author. There are several features that we can identify as involved in the documents, for example, the many styles juxtaposed in the same document, the sectioned short paragraphs (not necessarily linked), and the little use of traditional narratives (but without totally excluding them), reportage, essay, chaser and other literary resources, to our knowledge never regularly used in a military scientific document of this nature before, in other words the identification of a novel rhetorical architecture.

Although the documents were structured into unlinked sections, these sections gave more the impression of a union of ideas rather than of breaks in a single document (actually sometimes they cleverly overlap making the intersection of sections not empty). This worked very well for a wide variety of readers and numerous areas of analysis; however it is worth noting that in Blackett the "explanation", or even the use of metaphors, is generally related to geometrical images.

In this case metaphors are more explicative than comparative, as geometry is referred as a topological concept: space and the elements that span it but also constitute it too, in the abstract imaginative level. Note that this is not the *optical space* but the *geometrical space* thought in a clearly dynamical sense, as explicitly indicated in the following quotation:

> Expressed in geometrical language, the first task of Operational Research faced with a new operation is to investigate the shape of a multi-dimensional surface
>
> $$Y = F(x_1 \ldots x_n)$$
>
> surrounding the point corresponding to a past operation, and to use this knowledge to predict the properties at a neighbouring point corresponding to a future operation [Blackett, 1948, 182].

One must remark that gaining respectability within the experts' opinion was a central issue for somebody in Blackett's position. Of course the experts in war matters were the military but they were not to be influenced as if they were experts in 'science'. This does not mean that in other parts within the military body there were also scientific advisers[21] who could evaluate the recommendations coming from another section in the same military organisation.

In any case, the Operational research texts were of the highest impact in the military circles, possibly due to the con-text as the hermeneutical factor or the text *per se* as a new kind of discourse aimed to open new directions for scientists and executives in non-scientific organisations.

Blackett: A new writing Technique and a new epistemology for business logistics.

An important innovation in Blackett's work with the military, leads to the development of Operational Research in his innovative writing technique. He used it to translate to the military the 'scientific' methodology and results he used, for instance the use of ellipsis, i.e. omitted some words in the syntactic construction without loosing the

21 An example of this is Lord Cherwell, scientific adviser to the Prime Minister, Sir Henry Tizard or John D. Bernal, the last was scientific adviser in the Ministry of Supply, all of them working in other sectors and/or organizations sometimes independently from Blackett.

total meaning of the expression, (see [Blackett, 1948]). In his case, it is more a sudden leap from one topic to another letting the reader complete the ideas, in order to condensate the many subject matters involved. These texts treat problems using different approaches; for example statistics and differential coefficients; furthermore, in [Blackett, 1948] the use of mathematical resources is prominent.

If Blackett would have written nine or ten documents separately, each one treating a different topic from a different point of analysis, for example saying "from the statistical point of view the problem is seen like this..." or "using the method of differential variations the explanation is...", he would not have expected to encompass the difficulty of the problem, which is a main characteristic of his contribution to Operational Research in our view.

He does not keep his analysis by area of specialization but by problem, and that implied doing it in each single document. This is another feature characteristic of the Operational Research techniques reflected in the texts. And this is why we feel the literary skills mentioned above must not be omitted in an analysis of the subject. The nature of the documents forces scientists to dwell directly with the essential core of the matters considered, avoiding unnecessary information (that could be of the utmost importance in a scientific or technological paper) or deviating to ideas that in their circles could be presented as extremely relevant.

Aesthetically speaking the leaping of concepts is a difficult art; at the end of the reports one should be able to remember the beginning and purpose of the document from its first notes until the termination, so the nature and spirit of what was to be communicated is not lost. I will refer to this as the 'architectural structure' of his documents. In Blackett's case, texts' architectural structure is clear enough to enable the reader to follow the ideas till reaching the conclusion. This is managed first by the length of the document and the paragraphs that constituted it and, as said above, visually his layout was more of a melting together of sections and styles than their juxtaposition.

One can see cross-references from one section to another on various occasions as well as scientific concepts, a theorem or a management issue. This is a breakthrough compared with other contemporary styles of scientific writing texts such as in [Fisher, 1935], or for example purely administrative or purely military reportage. This point separates Blackett from other formalisers of the activities of Operational Research. Others might have mentioned the term ‘Operational Research’, or used some of the techniques used by Blackett, or cooperated with the military organisation before him, but we feel there is a substantial difference between Blackett’s innovative approach and that of a number of others in groups of Operational Research pioneers.

One has to understand that Operational Research has unique characteristics as detailed in the objectives of the Operational Research sections back in the 1940’s, and even more in the modern conception of the discipline. One of Blackett’s breakthroughs is reflected in the texts issued. Narrative was the style hitherto used by military intelligence (the organization in charge of gathering and analysing information) and it reflects the cultural background and their praxis. A good example is a report of Special Intelligence in Home Operations, analysing the sinking of the Bismark dated on 27th May 1941. We find the account with absolutely no numbers and concentrated on the narrative of the account although it has a foreword note: “As this history is written strictly from the point of view of intelligence and not from operations, it contains only so much account of the action as is necessary to explain the working of the Admiralty Operational Intelligence Centre (OIC))”:

> On the afternoon of the 20th, the Swedish cruiser ‘Gotland’ sighted the German squadron… It may be of interest to record in this connection that a German prisoner said afterwards that the ‘Bismark’ was sighted by the ‘Gotland’ off Bornholn in the forenoon of the previous day the 19th.
>
> Aircraft from ‘Victorious’ attacked with torpedoes about 0020/25th and made one hit which the c. in c. wrote was largely responsible for the ‘Bismark’ being finally brought to action. [PUBLIC RECORD OFFICE, ADM 223/88]

In the report of *Movements of German Major Units* from and to the Baltic and coastal movements between Norwegian ports from 1942 to 1943, there is nothing but accounts and narrations of special dates and cases, but no numbers or tables or graphics are to be found [PUBLIC RECORD OFFICE, ADM 223/88]. Although the objective is the same, to explain operations of the admiralty as Operational Research Sections were attempting to do. Of course the way to do it and the methods to achieve it made the difference and therefore made the works of people like Blackett to be regarded as a discontinuity in the field.

This is a qualitative difference in the way of working and analysing problems that clearly distinguishes the two groups and that illustrates very well the cultural differences and methodologies of the environment in which the discipline was developed for the first time.

Scientists like Blackett, Tizard or Bernal used to ask for data tables and this kind of narratives were not central to their work, as one of the first epistemic steps of Operational Research was to construct data out of the translation of the narratives into quantitative factors.

Another example can be found when Blackett was asked to prepare notes on a paper written by J.H. Godfrey, from the Directorate of Naval Intelligence:

> The two rival views at present under discussion are:
> That it shall continue to be used, in the main, for strategic night bombing and,
> That it shall be partly diverted to other operations such as anti U-boat, anti shipping and tactical support of land forces.
> Starting with the assumption that the result of (a) is extremely small, any or all of (b) becomes preferable....
>
> For the bombing offensive seemed the only conceivable way by which we might win the war, and since we are going to win it, it had to be by bombing. [Blackett Papers, Royal Society Archives, D64, D65]

The recommendation written by Godfrey contains no numbers, figures or graphs, and this was a main difference often disregarded by historians of the subject between the two sub-cultural but often so interpenetrated worlds above mentioned: those of the scientists and military.

Certainly, one of the first assertions that we must make is that Blackett's work is in the beginning (in intention) more reportage/essay than a scientific note. It is not certainly philosophical; actually it is more psychological in the sense of aiming to the communication-comprehension of theories and disciplines that led to the creation and development of Operational Research as an applied science.

Appendix of chapter 1.

The journals reviewed were: *Academy of Management Journal, Accounting Review, Administrative Science Quarterly, American Economic Review, Bell Journal of Economics and Management Science, Brookings Papers on Economic Activity, Canadian Journal of Economics, Econometrica, Economic History Review, Economic Journal, Economica, Industrial and Labor Relations Review, International Economic Review, Journal of Accounting Research, Journal of Business, Journal of Economic History, Journal of Economic Literature, Journal of Finance, Journal of Financial and Quantitative Analysis, Journal of Human Resources, Journal of Industrial Economics, Journal of Money, Credit and Banking, Journal of Political Economy, Journal of Risk and Insurance, , Oxford Economic Papers, Quarterly Journal of Economics, Review of Economic Studies, Review of Economics and Statistics, Journal of International Business Studies, Journal of Consumer Research, Journal of Occupational Behaviour, Strategic Management Journal, Managerial and Decision Economics, Journal of the Operational Research Society, MIS Quarterly, Canadian Journal of Economics and Political Science, Bell Journal of Economics, Academy of Management Review, Journal of Economic Abstracts, Contributions to Canadian Economics, Management Technology, Marketing Science, Journal of Labor Economics, Journal of Business of the University of Chicago, RAND Journal of Economics, Journal of Applied Econometrics, Review of Financial Studies, Journal of Organizational Behavior, Journal of Economic Perspectives, Journal of Insurance, Organization Science, Publications of the American Economic Association, Brookings Papers on Economic Activity. Microeconomics, Operational Research Quarterly (1950-1952), OR, University Journal of Business, American Economic Association Quarterly, Journal of the Academy of Management, Journal of the American Association of University Teachers of Insurance*, and *Proceedings of the Annual Meeting (American Association of University Teachers of Insurance)* compared with 73 in mathematcal journals. The mathematical journals searched are: *American Journal of Mathematics, American Mathematical Monthly, Proceedings of the American Mathematical*

Society, Transactions of the American Mathematical Society, Annals of Mathematics, Econometrica, Journal of Symbolic Logic, Mathematics of Computation, SIAM Journal on Applied Mathematics, SIAM Journal on Numerical Analysis, SIAM Review, Philosophical Transactions of the Royal Society of London. Series A, Mathematical and Physical Sciences, Proceedings of the Royal Society of London. Series A, Mathematical and Physical Sciences, Philosophical Transactions (1683-1775), Philosophical Transactions of the Royal Society of London, Philosophical Transactions of the Royal Society of London. A, Philosophical Transactions of the Royal Society of London. Series A, Containing Papers of a Mathematical or Physical Character, Abstracts of the Papers Communicated to the Royal Society of London, Abstracts of the Papers Printed in the Philosophical Transactions of the Royal Society of London, Journal of the Society for Industrial and Applied Mathematics, Proceedings of the Royal Society of London, Philosophical Transactions (1665-1678), Analyst, Journal of the Society for Industrial and Applied Mathematics: Series B, Numerical Analysis, Mathematical Tables and Other Aids to Computation, Journal of the American Mathematical Society, Proceedings of the Royal Society of London. Series A, Containing Papers of a Mathematical and Physical Character, Philosophical Transactions: Physical Sciences and Engineering, Proceedings: Mathematical and Physical Sciences, Proceedings: Mathematical, Physical and Engineering Sciences, and Philosophical Transactions: Mathematical, Physical and Engineering Sciences

Chapter 2
PMS Blackett and the Emergence of quantitative business analysis: a semiotic approach.

Introduction.

In this chapter I will show that military were aware of the application of scientific methods to war before Blackett and the "emergence of Operational Research as business analysis" as such during the Second World War, and relate this state of affairs with the mathematics developed by Blackett in the 1930s and 1940s. I also analyze, in a semiotic way, concepts and their utterance in Blackett's discourse on the application of science to war (or any resource management planning). This analysis is followed by an epistemological discussion of Blackett's work in the early years of operational research, where quantitative analysis was a mere exploration.

As seen in the previous chapter, the rather academic discussion on whether Blackett was, or was not, the "father" or the creator of Operational Research (in his role of consultant to the military) the latter understood as the application of science to military affairs, is often linked to the question on whether before him military had been applying scientific concepts to war. Arguments such as: there were some people before Blackett who used some of the language or vocabulary that Blackett used referring to Operational Research activities or that there were groups within the military organisation that had a similar agenda of that of Operational Research Sections (like the Tizard Committee) in any case, are incomplete in the sense of the generation of knowledge; in this case, one has to go back further in time looking for similarities and "pre-histories" of Operational Research to address properly an argument such as the exposed here.

We might refer to political and military theorists as well as scientists acknowledging the importance of scientific concepts in the practicality of war affairs, as pointed by Neal Wood in the introduction of [Machiavelli, 1965][1] that Mathematics, of course, had being applied increasingly to the problems of military engineering and artillery. Moreover, statistics in the form of tactical tables represented the application of mathematics to the organisation and control of human beings. The use of quantitative methods for the planning and direction of troop formations and manoeuvres presupposed an emphasis in theory and practice upon discipline and drill.

It seems that one of the first men who "revolutionised" warfare while applying mathematics to questions of military organisation, was the astronomer and mathematician, Thomas Digges[2]. One result of his military experience was the publication in 1590 of an arithmetical warlike treatise named Stratioticos, the revised edition of a book which first appeared in 1579.

In order to get some perspective, in this paper I will quote the work of 1590 of a Spaniard called Andrada which makes reference to the importance of mathematics in war matters, as an example of what some might call a "pre-history" of Operational Research, I do not claim, however, that this work was as 'inspiring' as Blackett's or other pioneers' work. On the contrary, Blackett and his so-called "circus" had the importance that no one in the past had: what he created (the texts that led to the development of the discipline) overleaped the application and the incorporation of mathematics to practical life, and the immediate expansion of the discipline that covered what later was known as the scope of social sciences.

1 Originally published in 1521.

2 Thomas Digges, (1546–1595) As well as having a military career, Digges also wrote and worked on other military matters. His book Stratioticos (1579) is a mathematics book for soldiers and contains the first discussion of ballistics in a work published in England. He also worked on fortifications, being in charge of the fortification of Dover harbour in 1582. A year earlier he had been involved in producing plans for Dover castle. Digges was a member of parliament from 1572 and again in 1584. His military career was with the English forces in the Netherlands from 1586 to 1594. In 1579 Digges wrote his military work Stratioticos which he dedicated to Robert Dudley, Earl of Leicester. Dudley was named governor-general of the Netherlands in 1586 and Dudley appointed Digges to be master-general of his army to assist him in the campaign.

A number of circumstantial reasons may justify why Blackett had the impact that others (even contemporaries) did not have[3]. Nevertheless, in this chapter I will apply semiotic tools to try to discuss directly on the textual work of the main characters the emergence or the creation of Operational Research as a discipline in mathematics. This approach contemplates sociological, political and scientific factors, but does not bias the approach to any of those, but remains in the pure semiotic sphere, and therefore does not deviate its attention on other surrounding matters that, undoubtedly, are important, but not centrally unique in the history of science or technology, for example making social history of Operational Research would leave a gap in the scientific or technical sides of the events that, in the case of our concern, gave birth to this new branch of mathematics.

One of the main concerns of the History of Science is the generation of knowledge in science[4] and one of its basic tools and resource of study are the texts (as primary sources), therefore, a semiotic approach in the sense of textual analysis can give us a wide and quick scope and first analysis of how knowledge (and its context) in its textual form flows from one state to another. Classical Semiotics contemplates three parts of signs: 1) the utterance of the sign, that is, its form or in de Saussure's terms the signifiant, (see [de Saussure, 1966, 65]) 2) the concept to which that significant is referred, or in de Saussure terms, the signifié, and 3) the object to which the significant is linked to. This triad has been termed differently by other author, for example, in Peirce's notation as "Representamen, Interpretant and Object" (see [Peirce, 1986, Vol II-274]).

3 One can appeal to sociological, political, scientific or philosophical arguments.

4 Sometimes authors centre their research on other factors, to the point that it seems that science is only a mere excuse to do biography, social or political history. If their object of study (as some say) was the social factors of science, then our discussion would turn philosophical and we should question deeply on the object of study of the History of Science and Technology, in this case of social history the focus of the research becomes blurry and science or scientific discovery become only another variable of the universal sociological system in which science and everything else that exist are submitted to the sociological explanations.

In order to understand the representamens or the utterances of signs, it is important to analyse the words used in the respective speech; to comprehend the interpretant it is crucial to have a view of the cultural or social environment in which the sign is located. Therefore in this chapter I analyse on the one hand, a glossary of the most important terms in Blackett's discourse (i.e. his signifiant), and on the other the military background in which Operational research was developed at that time (his signifié). In that way, objects of reality, texts and the reception of these texts can be analysed in a thorough way.

Semiotic analysis can provide historiographic explanations and can offer an ordered historical-sociological ground. However, having a mere semiotic analysis can be complemented with other approaches such as the hermeneutical as semiotics only gives us a general scope or a background on the generation of knowledge but has its limitations in further cognitive analysis; in order to continue understanding this generation of scientific knowledge one has to apply other tools that are better able to explain it. As said before, understanding this generation is one of the objectives of historical enquiries. Consequently in this paper I have included, as well, the connection of semiotics with an epistemic (or, I would say, post-epistemic) level.

Writings about the military before the age of Operational Research help us also to construct a comparative platform to analyse Blackett's breakthrough in this area. One has to recall that Blackett was also trained within the military, during First World War, at a time when persons like Colonel F.N. Maude (see [Pasley, 1914]) wrote about similar issues referring to them as the "Art of War".

The way in which Blackett's thought was focused is reflected in his work, and his work is reflected in his writings and papers. So analysing them from an epistemic point of view, i.e. dividing in some way the steps in which we can locate a generation of knowledge, may help understanding Blackett's position and aims in the micro-sociology of the military organisation.

Studies of war in those times relied strongly on the historical point of view; the statistical or numerical approach was acknowledged but not widely applied and accepted. Moreover, the concepts assigned to words were not necessarily identical to the concepts we now use. For example, a concept of mathematics, that referred mostly to arithmetical tasks, changed through the close interaction between scientists and the military (into a more geometric approach), and this kind of facts are a substantial part of Blackett's breakthrough in the development of Operational Research as a coherent chapter of science.

Some remarks on art, science, militia and management before Blackett's times.

Let me return briefly to the question of the war in which military organization analysed operations for centuries, before the Second World War.

In a document, presumably from 1570–1590, a Spaniard Manoel de Andrada Castel Blanco, sent Phillip, King of Spain and Portugal, a document in which he develops a strategy to defend his empire and the faith[5]. In the introduction he begins explaining shallowly the methodology and principles of the analysis achieved in the book[6].

Words like "mathematically deduced", which would have meant a different thing nowadays are often mentioned in the book; he also

5 The original title is: Instruccion que a V. Magestad se da para mandar fortificar el mar Oceano y defender de todos los contrarios Piratas, ansi Franceses, como Ingleses, en todas las navegaciones de su Real Corona, dentro de los Tropicos".

6 "First it must be taken into consideration that the whole navigable Ocean Sea belongs to your majesty, in all the directions of the four winds, and that rules legitimately all the Reigns and conquests, being defender of the Faith, Being this reason the cause that the pagan English and French rob in all the Ocean Sea in all its shipping, from the east as well as from the west, becoming rich and powerful, to maintain wars against the Catholic Church and against your majesty's crown. For the defence of that, I shall write down all the remedies that below will be stated, that they are mathematically deduced, respecting the dimensions of the world of the Quadripartito of the rising and setting of the sun and the movement of sea tides, which by their nature have their rare and varied effects" [Andrada, 1590, Introduction].

associates "mathematics" with useful tools for war and navigation'. Historically it is of some importance to ask ourselves what did they (Andrada, Blackett, the military, etc.) mean by these side references, for example "mathematically deduced"; because in this analysis we can introspect the theories and ideas in a cultural context; that is, in this way we can approach the proper scientific ideas as a main subject of study, without dismissing sociological factors and other components.

This idea is clearer if one puts in perspective that culturally (and here begins our semiotic analysis) the word mathematics meant, or referred to different objects in such a way that the person who uttered the word (generally a scientist) and the person who read it (in ouir case the military) diverged in communication, i.e. interpreted different things and therefore in interaction, when printed in a text, this is why the relevance of past interactions of scientists and military were not as effective as in Blackett's case[7] or perhaps, were not recorded as intensively.

For example, historians that follow Kuhn's ideas are, sometimes, partial in this respect, if one tries to define "normal science" in those times, one will find that either, it is very difficult or one has to become an officialist of history; this means to accept that the sociological, and I would say institutional part of history is the main conductor of ideas. This may be admissible in some cases, but does not seem sufficient water tight to justify that such description of facts can be taken as a explanation of ideas nor science.

In this sense Semiotics offers some useful insights as it gives the approach of the text, its elements and their meanings in a given context and not the other way around, i.e. in which the context without the text can be anything in which we might get lost easily, for example analysing the impact of an earthquake in Tasmania on the scientific discovery of Blackett.

The classical triadic scheme of semiotics (sign-object-interpretant) explains in a global manner the textual utterance, the ideas and

7 As [Eco, 1976] says, the reader always read with a dictionary and with an encyclopaedia; one gives the word the main reference to an object and the second the context in which that word must be interpreted.

concepts to which it is linked and the context of reference in which it is uttered.

In this sense we can analyse another important fact in Andrada's book: some might argue that Andrada in the 16th century had a similar approach and even the same kind of speech that Blackett four centuries later, and one can infer in consequence that Andrada might not have had the detailed and correct sociological conditions to launch the idea of the application of "scientific analysis" to war to military commands[8] and therefore begin a sociological enquiry on the topic, nevertheless, semiotic analysis can throw light on these matters specially in the sociological spheres without losing relevant meanings according to the use of signs.

In this sense (sociological), the interpretant can be analysed in the following way: along history, the military had its own methods of developing strategies some of them were through the historical analysis of the past great generals and their famous battles, in Eco's terms, the encyclopaedia with which the reader interprets (In Eco's thought, a reader (lettore in fibula) reads with a dictionary and an encyclopaedia which gives him a referential definition and a context for the meaning) pointed that mathematics was a formative subject and an area of specialised technicians under the orders of an expert in war, and it is in a similar context where science as conceived and introduced by Blackett, got in the scene, and not in the case of Andrada.

A relevant case, which is near to Blackett's in the semiotic sense, is the fact that scientific applications were mentioned in military lectures as part of the military training in Britain, where the military had a policy on science collaborating in war subjects, as stated in a lecture delivered to the Organization Society in July 1912, only two years before the outbreak of the First World War, by Colonel F.N. Maude, C.B[9].

8 And it was with the art of Mathematics that Francis Drake became such a good navigator and such a good corsair, that in the years of 76 and 77...going with seven galleons robbing and destroying the Spanish shipping over almost the whole sea as far as Magellan Strait... [Andrada, 1590, Chapter 4]

The lecture begins with some statements about what he calls "Science of Organization", considered as an applied science. He is concerned with the study of war strategies, which he later calls 'The Art of War': "The Science of Organization is a synthetic science the principles of which are involved in all the applied sciences. War as one of the arts or applied sciences, which as Cicero says, are connected by a common link or blood-relationship to one another, is closely affected by the principles of this Science. The Science of Organization forms in fact the basis of the Art of War, and incidentally shows the connection of the Art of War with all other human arts". [Pasley[10], 1914, 147].

For him war activities are connected to science and to human arts. The interpretant is given by the historiographical account of the battles , for example, an important note in his lecture is that, he analyses the case of the Russian Campaign of 1812 in the history of the French Army. In his words Napoleon knew very well the secrets of this "Art of War", and considered that the greatest of all the secrets was the 'economy of effort'. He analysed for example, the question of communication inside the Army, that is, how orders reached the battalions from the generals down. In the old Armies orders had been issued from Army Head Quarters direct to the regiments, battalions, and so forth; the consequence was an incredible waste of time in copying and recopying orders.

Putting it in modern terms, a problem that Napoleon was concerned with was the logistics with which orders flew from the Headquarters to the regiments in the battlefield; he considered 'ordering time' as a main variable of the war strategy. Napoleon's

9 Col. Frederic Natusch Maude, was born in 1854, died in 1933, entered the Royal Engineers in 1873 and became Staff of the College Graduate in 1891, he has many publications, some, for example, Evolution of Strategy attracted considerable attention and more even translated into German. He also wrote War and the world's life, Campaign of Leipzig, of Jena, of Ulm and many other technical essays.

10 Sir Charles William Pasley (1780–1861) was a General, Colonel Commandant of Royal Engineers. In November 1810 Pasley published the first edition of his 'Essay on the Military Policy and Institutions of the British Empire'. It attracted great attention and ran through four editions.

orders reached the men in from 3 or 4 hours, whilst in the armies of the Allies in 1813, it took them 12 or even 18 hours before the troops actually moved in response to them. Hence the men wasted all day waiting for orders, and then had to march all night to reach their appointed places, thus going into action worn out from out of proper rest and refreshment. (see [Pasley, 1914, 154])

According to Pasley, this 'ordering network' was an opportunity area which Napoleon noticed, and therefore improved.

Semiotically the reader of this lecture can establish the "Art of War" as directly connected to science, moreover mentioning concepts such as "synthetic science" (which was mentioned in his lecture) in relation with the Napoleon's problem as analysed and exposed in this lecture. An important factor readable in this lecture is the concern of military on the speed with which war was conducted. In Napoleon times, the technological advances limited speed and did not allow managing a large number of variables at the same time. A first problem in the management of resources was that when soldiers had reached the appointed-place they were to have some proper rest and refreshment. Since those times war was conducted in a certain level of abstraction, for example on maps and tables, which in the general were studied in headquarters, if one analyses this situation (difficulty of communication) one can see that there was not much place and time for detailed orders once that an initial order was given. Problems in telecommunications as others were reflected in this analysis (up to 18 hours for reaching orders in the armies of the Allies), so we can see that military naturally were not strangers to this kind of "optimising" approach in the strategies and tactics applied in the field.

Years later, in the Second World War the availability of technological advances made the activity of war far more complex, detailed, technical and specialized to the point that variables such as visibility, or location of enemy's facilities became more and more relevant to the problem. Furthermore, the military began to realise that all of these variables acted at the same time and not all had the same weight as others, to the point that discussions on whether the morale or the psychological state of the enemy became as important as new warfare, the number of escorts for vessels or the implementation of strategic evasive manoeuvres of the battalions.

War operations became a very complex problem of strategy not anymore treatable only through a simple arithmetic rule or via historical analysis of campaigns, but addressable to scientists.

Although there was some arithmetical calculation and a quantitative evaluation, the "scientific" approach was not thorough; the formalisation of the scientific activity within the military was not completed, i.e. there was not a scientist as such in charge of these matters and sometimes the approach was still more historical, for example the study of the wars of Frederic the Great, was among the topics analysed in detail in military academies.

In general there has been several books where military developed their strategies and tactics, most of them not based in a "proper" scientific source but rather psychological or based on experiences in the battlefield, as in the "Reveries or memoirs upon the Art of War" by Field-Marshall Count Saxe, dated in 1757.

Up to 1914, one can found that scientific approaches were used in war activities but the attempt to transform war into a scientific study was not yet well received. As warned in his lecture: "In fact to make the Science of War a geometrical study is to rob it of its most important element, its soul or psychology" [Pasley, 1914, 157]. It was then recognized that war was a matter depending on a large number of variables, nevertheless some weight was given to the variables more difficult to quantify, such as, the morale or the fatigue of soldiers.

Although before Blackett and his team some scientific approaches had been used to develop strategies for war, it was during the Second World War when one can detect a novel approach regarding the collaboration of military and scientists. This collaboration was based in a communication, however this communication, as a "Great Narrative" has as a principle, the fact that the one that is addressing means what he says, the belief that there is an adjustment between word and meaning is not given, but is achieved in a process, in a construction that is updated and modified (see [Glasersfeld, 1995]). That is why in the iteration of the communicative process Blackett achieved a great success.

Blackett and his circus: *mise en scene* of Operational Research as mathematics applied to business analysis.

To lucubrate on the reasons why Blackett's approach, especially concerning the collaboration of scientists with military during the Second World War, in general was successful[11], might be as well analysed from a textual point of view as we have seen in the above section. Textual analysis is a tool with which we can better understand the generation of knowledge in this specific area. The reader might ask why instead of basing the argumentation on one of the, undoubtedly, most influential books of the 20th century, The structure of Scientific Revolutions, I prefer to base it on textual theories, the answer lies in the idea that establishing a paradigm, which in the case is in the middle of the Military sphere and the scientific tradition of physics, was inexistent. In fact any paradigm with respect to science lied in the historical revisionism and its continuous repetition via the reprinted articles in official publications. As [Fuller, 2000, 318] says, "The science studies community currently suffers from self-inflicted Kuhnification. The main symptoms are a collective sense of historical amnesia and political inertia, which together define a syndrome, call it "paradigmitis".

In this sense if we use the concept of paradigm, then normal science should be understood as an esoteric sect, crede ut intellegas ("believe to understand"), the question is that this esoteric happens to be the "official" sect. As put by Feyerabend, the main danger of Kuhnian proposal was reducing the scientific education to an academic and bureaucratic training designed to extend strategically a status quo as long as possible. In the case of Blackett and the military intelligence this was impossible, for obvious reasons.

The question of scientific knowledge can be, following some authors like Feyerabend, Lakatos, Popper or Serres, of heuristic imagination, theoretical creativity, rational lucidity or speculative and historical opportunity, but definitely not just as an administrative

11 To the point of developing new branches of mathematics and areas of study and specialization.

problem, seen as an execution of policies and university control, as the Kuhnian theory might lead to establish. Therefore ideas supported by the previous ones might fit better to understand the proper cognitive process in Operational Research.

For example, for Popper to "make a discovery" consisted in identifying epistemological limits of a determined theory which was bounded by a intuition of reality, for Kuhn, any discovery is a academic activity in the sense of publications and established journals, that is in a very especial way spread and narrated. Then an idea like: discovery does not have to be something new or a "Eureka experience" is an argument that not many academics can overcome or understand. This kind of academics is only looking for a specific bureaucratic efficiency of the Kuhnian proposal, i.e. "commitment to a specific scientific paradigm". And therefore, as Fuller puts it: "What philosophers of science today call 'methodology' turns out to be a secularised version of the project of troubleshooting the sources of error in belief formation, until we are left with only one plausible explanation". [Fuller, 2003: 115]. Blackett never intended, and never published with that aim, moreover, the training I Operational Research Sections was internal and methodologies were free to be built as suited the problems.

This conception of scientific methodology is widely treated by [Latour, 1993] who considers the laboratory as epistemic centre of modern science. For Feyerabend [1981a], the laboratory is a place to take decisions that can be changed by others, for Popper, it is a place where hypotheses and conjectures are to be experimented as many times as necessary, ideas closer to the case that we are treating in this paper, but not in the Kuhnian sense in which one would have to assume that the laboratory was a place to confirm the beliefs of scientists to justify their notion of reality given by a certain paradigm. Under the Kuhnian perspective, experimentation does not generate doubts nor generates new perspectives or question the established theories, i.e. is not epistemic, on the contrary, experimentation only works for the legitimisation of the current normal science.

Considering the sociological argument of scientists collaborating with military, Blackett had also a breakthrough; this was achieved by the establishment of a close relationship in which scientist and military worked together at various different levels. This argument can be followed by the same texts and its semiotic analysis. In the following memo classified as "OS.10956" dated on the 31st October 1941, the aims and purposes of the Operational Research Sections are detailed:

> Operational Research Sections have been established at Headquarters of the Bomber, Fighter and Coastal Command. The purpose of these sections is to provide Commanders in chief with trained scientific staff who can collect data and undertake research into an analysis of technical, tactical and general operational factors from which can be drawn deductions and lessons which will serve to guide C-in-C and their staffs in the conduct of future operations and in formulating the operational requirements. [Public Record Office, AIR2/5352]

In this way, Blackett and other pioneers of Operational Research had the sociological, institutional and administrative background which now seems to have been necessary to develop their work. It can be understandable that the administrative discussion and the delimitation of spaces continued during the war years, as seen in a letter dated on the 4th September 1941, from Sir Henry Tizard to C.E.H. Medhurst , where he stated that professor Bernal who was scientific adviser to the Ministry of Home Security had been examining broadly the efficiency of enemy night bombing in England. Nonetheless, they discussed who was responsible for endeavouring to assess from all possible sources of information the percentage of English bombs falling on urban targets on Germany and who was responsible of examining and advising on the destructive efficiency of particular types of bombs and thereby guiding technical policy. This is where Blackett and the Operational Research Sections had their main work.

Text in military circles.

Blackett's texts are situated in the environment mentioned above and can be divided, for their study in at least the following five elements: 1) essay, for example section one of [Blackett, 1948] that shows what, in his point of view, the scientific work within the military service was; 2) scientific report, as in section two subsection ii) of [Blackett, 1948], which mentions the use and quotes, the mathematical formula of the Poisson distribution; 3) representation by anecdotes of real problems, for example, the same section (two) of [Blackett, 1948] and section five which make allusion to the "top hat fallacy" ; 4) narrative, for example, the whole of section two of [Blackett, 1948] and; 5) the autobiographical essay as in [Blackett 1962].

These elements coexist with each other with exception of the autobiography. They clarify and explain each other as they explore a single theme like an accordion that squeezes in the theory and scientific reports, and releases air in the anecdotes and essays illustrated with examples that, in the context of an essay and scientific report, are less examples and more situations worthy of special attention and careful analysis.

This kind of semiotic analysis brings in an explanation and an interpretation, in this case, in the context of a war operation in the military service, these factors embrace them simultaneously; but this only happens at the moment they become part of a document, opinions change their essence to a formalised opinion supported by a scientific argument and the utterance of signs become more permanent as they legitimised the opinion.

Outside this written context the persons were in a territory of assertion; everyone held his truth via his statements: i.e. the minister, the scientist, or the officer's biological dimension. Within the space of Operational Research texts, no one asserts unless the proper mathematical procedure was followed: in a way, it is the area of observation, hypotheses and proving. In Operational Research, assertion is essentially inquiring hypothetically. That is, where the discussion on the difference between notation, denotation and connotation begins.

This Accordion mentioned above, also works for the particular and general statements. The first represented by specific cases and the latter by the physical or mathematical models themselves. In this process, words, numbers and graphics are used and work together in the accordion of scientific-colloquial worlds. In formal semiotic terms I can refer to [Eco, 1976] where he states that a reader has always a dictionary and an encyclopaedia, where concepts are stretched in a similar way that I mentioned above (see footnote page 6).

From two poles: the specific case and the generalization, the scientist in Operational Research generated tokens, that were not purely scientific, but were effective in creating a psychological imagery in the counterpart-reader.

These tokens were of three kinds that were correlated to the object of their meaning. In the following figure we can look how the rhetoric of example-theory worked in the generation of tokens.

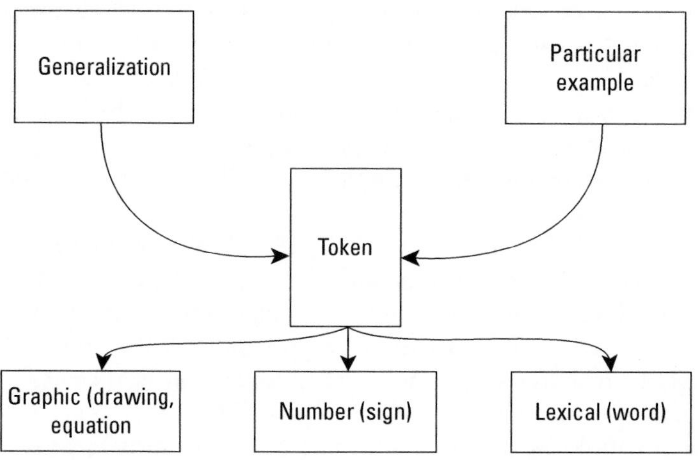

In this context, a dogmatic (or in Kuhnian terms paradigmatic) thought turns hypothetical, i.e. there is not a founded idea on the part that is receiving the theory (the military), and in this semiotic sense, Popperian theories on the construction of science can be applied, i.e. the proposal and revision of hypotheses, not like the article of [Eilon, 1975], pretending that Operational Research had a unique scientific method, (as if there existed one).

The idea of a double discourse is crucial to understand relevant issues in the reception of Operational Research reports; that there are two languages, the language of the executives and the language of the scientists; one refers to the acts and the other to the ideas. One is more phenomenological and the other is more idealistic, therefore Blackett builds an explanation for the scientists (for himself and for his group) and an explanation for the military.

At the end, a new semantic and a new set of meanings and words were created based on the experience and activities of Operational Research sections. This language is the most patent proof of the transcendence of Blackett's work in this field.

This analysis begins to turn more complex, because one has to generate an epistemic explanation to understand passing through the chiaroscuro from the idea to the action (idealistic to phenomenological and vice versa); also to analyse this flow semiotics is useful: Blackett's discourse can be classified as a writing activity in which the use of images melts with narrative creating what we can call a visual literature.

Image is a common factor in the topics discussed by Blackett and of the documents themselves; it appears several times over Blackett's course of explanation and helps the "accordion" work smoothly, for instance:

A problem is on the one hand, the phenomenological inquiry (implementation, action) and on the other the ideological investigation. Operational Research is the vehicle to evaluate and, more important, to justify the good actions (or the best possible) via the application of analysis to real life situations. This investigation is revealed in the examination of certain words that are associated with images on one side and with ideas on the other, some of these words are: geometry, experiment, experience, quantitative, qualitative, probability, deterministic, and system.

In the development of the documents these main words are interpreted, analyzed, studied, defined and redefined with different notations, denotations and connotations, and then, via the generation of images, changed into icons of expression. In this sense there is an image (a product of perception) that has a direct equivalent (or translation into words). Words are referred to images in the following sense: names have their equivalent in perceived objects (bodies), adjectives to qualitative properties of the perceptions and verbs to actions and reactions of images among themselves; see [Deleuze, 1983, 59].

Then the documents are built on these bases and within this frame, for instance in [Blackett, 1948], the basic word-images are: scientific, phenomena, observation, experiment, methods, function, coefficients, statistical, causal, theorem, explanation, and solution.

Already in the word-image level, there are two concepts that I want to remark concerning the semiotic analysis under this epistemic frame: the syntax and semantics. I refer to syntax as the logical rules that allow a system of signs to work, i.e. its structure. Semantics is concerned with the values that a sign could have in certain syntactic context. Notation is referred to "Graphia" or the visual image that signs have, denotation is the significance that a sign has as referent to a given system of values or things, and connotation is the value that the referent has in a cultural environment given. The problem is easy if we refer to the word dog, this word (with its many notations like hund, chien, or perro, connotes the animal mammal, quadruped that barks, and connotes fidelity or protection. Although in cases with words like "domestic" the problem gets more complex. In general, the cultural environment is the one that connoted and the reference is the one connoting. In the case of scientific concepts like those treated in Blackett's documents, for example "probability", denotation and connotation are taken, managed, flown and adopted on both sides of the communication scheme (scientific-military). In this case, sometimes we have the opposite situation: the cultural environment is the one that denotes and the reference is the one connoting.

Blackett's glossary for executive levels: the key of communicating effectively.

Geometry, has won a richer meaning, probably it is the most important word in Blackett's glossary, from Euclidean to the topological one. It goes from the proportion of figures to the study of the shape of some forms (even of more than 3 dimensions). Speaking of a modern approach to geometry and differentiable functions acting in geometric objects, one has to speak about topology. The same Poincaré defines the spirit of the discipline, [Poincare, 1895–1928, 194][12].

In Poincaré's Analysis Situs it is stated that his subject is more a qualitative than a quantitative study. This idea was not only transcendental in mathematics, but in physics too, models that were used to represent natural phenomena assumed some principles, some of them philosophical other mathematical, such as if the universe is or not deterministic or if the space was continuous or discrete, such assumptions were developed for physics and the so-called exact sciences in the beginning of the 20th century. This kind of assumptions were made by Blackett specially due to his formation in experimental physics, and some of these concepts are reflected in Blackett's work of Operational Research, for example assuming that the operations of war (a human activity) could be modelled by a smooth continuous function and that one could build a valid partial differential coefficient to measure defence activities.

This kind of ideas was followed not only by mathematicians, but by physicists too. Modelling was not anymore just the task of solving a differential equation, but also had the objective of characterizing the space in which the events happened, this means, to see the natural phenomena in another different way:

Events in nature were not events anymore, they became only the meaningful observable parts of them, (that is the invariants) and further more, they became the parameters of the observable parts that could be perceived (images again but this time in the form of a mathematical sign), that is, entries of vectors in a space that represented the meaningful objects of physical phenomena, and of course, their asymptotical behaviour. This kind of models could not be possible without concepts like the concept of probability and the definition of Manifold. In Operational Research discourse there is a constant reference to the operations as functions of several variables and their correlation. In this sense this scientific minimalism is a crucial characteristic of the discipline.

12 Geometry is the art of reasoning well with badly made figures. Yes, without doubt, but with one condition. The proportions of the figures might be grossly altered but their elements must not be interchanged and must conserve their relative situation. In other terms, one does not worry about quantitative properties, but one must respect the qualitative properties that is to say precisely those which are the concern of Analysis Situs.

As said above, to build a mathematical model with these tools one has to make some assumptions, such as continuity or smoothness, (that is if the function is continuously differentiable). In Blackett's case; we can see a good approach to model the qualitative properties of the problems. His language includes mathematical and specifically geometrical concepts widely applied in physics.

The way that the concepts related to geometrical analysis could have been received as the schemes or graphic representation of strategies, as the military tradition was used to, for example, in the military training it was known that Frederick the Great never used the word strategy, for it was not a part of his vocabulary. Although the concept that later writers defined as "strategy" existed in the eighteenth century, the term used then by most military men was campaign plans (projets de campaigne); indeed, the two terms were used well into the Victorian era. Initially strategy meant merely "the science of military movement beyond the visual circle of the enemy or out of cannon shot" [Luvaas, 1966, 306].

Therefore scientific concepts concerning the study of entities that could not be seen became clearer reading the kind of documents that Blackett wrote. Being aware that strategy dealt mainly with things beyond the visual circle gave grounds to an effective reception of Blackett's scientific discourse.

Scientific, which, from that moment on, had more special meanings; the militaries owned a meaning of the technical physics, mathematics or other "traditional' sciences. This term is associated with the idea of dealing with facts in a quantitative way:

> This is analysed as quantitatively as possible and the results achieved are explained in the scientific sense, i.e. brought into numerical relation with the other numerical facts and the known performance of the weapons used [Blackett, 1948].

This quote is of the utmost importance because we can see Blackett's definition of science which happens to be relation-like. In the practical side, it is this "numerical relation" that gave a parameter to measure in some way, the options for decisions on operations and this was the scientific sense. Numerical analysis, due to the technological tools available in that time, was a difficult task that needed more of collaboration between some functionaries in the military structure than the isolated work of an individual. According to Blackett the role of calculation was work for mathematicians but not for the Operational Research team: The "Air Warfare Analysis Section" (AWAS) was a section with mathematicians working in it dedicated to make calculations.

Blackett needed their support for that, but not for included them into his section. In the 1st meeting of the Operational Research Committee, Blackett suggested that the AWAS would have a true function that was to act as a pool of specialist mathematicians for high grade mathematical analysis which was outside of the scope of the staff normally available at the ORS's and DSR, all (the present in that meeting) agreed that the original function of the AWAS as formulated by Blackett was to act as mathematical consultants, that was to make arithmetical calculations [Public record Office, AIR 2/5352].

It seems that Blackett saw the "geometry" as closer to the qualitative and the "scientific" more closely assorted with the quantitative.

Methods. It is a word of action, this word is never supported, never specified, never analysed. It is more an act of faith that the method in its character of transformation has an irrational input that gives a rational output, a graspable and readable result.

For example, in 1940 J.D. Bernal issued a report entitled "Principles of Shelter Policy" that was taken by Blackett to support other of his documents:

> The main object of the Research and Experiments Branch is to provide the basis for an effective and rational air raid precaution policy.
>
> For all shelters not proof against direct hits, this probability is 1 or certainty for all bombs, within a certain distance of the shelter it is 0 or certain safety for all bombs beyond a greater distance. Between these limits it has intermediate values. If we can determine either by experiments or experience what the values of these distances or probabilities are, it is possible to sum the probabilities over the area round the shelter so as to arrive at a kind of average area characteristic of the shelter and the type of bomb used.
>
> It is useful in assessing the relative value of a shelter to have another method of indication more easily grasped than the absolute danger area. This is the "relative degree of protection". If we wish to compare a shelter of vulnerable area A with one larger vulnerable area S as standard, we can say that A has a percentage degree of protection given by 100(S-A)/S. [PRO, HO 205/179]

"It is useful in assessing the relative value of a shelter to have another method of identification" he did not specify a method even before, not steps, or any other network, the explanation as many of Blackett's in all the Operational Research rhetoric, comes from comparisons and is aimed to specific results and goals. The issue that mattered was the final output that came from that method. Many of the statements in this area reflect the urge for administrative effectiveness as well as operations effectiveness, for example the initial paragraph above establishing the objectives.

Function, takes several meanings, the first is the mathematical meaning a relationship between elements from one domain and another. And the other is actions and reactions among the different parts of some system, for example the function of some institution in society v.g. marriage could be seen as a social institution with some functions associated. In this case, the nomination of function in a colloquial (non-technical) way is referred sometimes to a transformation, other to a relationship and other times to an action.

On the military side the socio-semiotic analysis—as [Latour, 1993, 83] would name it—comes near to the concept of a war operation, and these kind of contact points, sometimes tangential, could have been the reason why Blackett and his circus were so effective and transcendental in the implementation of analytic groups in the military and its further development into other areas of society and "normal life" of people. As explained by [Pasley, 1914], technically, the word battle is applied only to the decisive struggle, which decides the result of each strategic operation, and each strategic operation is conditioned by the numbers available, the pace at which they can move and the intelligence directing the moves.

This was understood by the military as it was reflected in their instruction: "In order to enable the student to visualize the interdependence of these opposing forces [the pace at which they can move and the intelligence directing the moves], we make use of the symbol of a serpent, depicted as pursuing his sinuous track through the obstacles environment opposes to his progress [Pasley, 1914, 148].

The convergence comes on two sides: the geometric representation of the objects of war operations and the conceptualisation of a war operation as a function of the variables available in a numerical form.

Theorem, this word lies in a different system of values, not precisely the mathematical system, beginning with assumptions like "In an intelligently controlled operation of war", "reasonable tactical latitude allowed to the participants" or "exceptions will occur for various reasons". This word connotes more a human condition than a logical derivation of steps.

Explanation and solution, are in some ways synonyms in Blackett's vocabulary, referring to the output of the process in which scientists analysed operations.

Experiment vs. Experience. There is an essential difference concerning these two terms: An experience is made out of observations, this is worked by Blackett in his [1948]:

> Experience over many parts of our war effort has shown that such analysis can be of the utmost value, and the lack of such analysis can be disastrous, [Blackett, 1948].

But further in the document there is a section entitled "Operational Experiments", where he refers to the gathering of data; does this mean that an experiment for Blackett was much of recording the observations of some phenomena? The answer is no, not only because he divides this concept into two main activities:

- Mathematical Calculation, and
- Test by actual operations, [Blackett, 1948]

In this case the visual set of inscriptions is also a language in the sense of expression. This is supported ahead in the document in a section of operational requirements he mentions:

> Nothing in this section or in section 2 (the section "Operational experiments") should be taken as implying that an Operational Research Section should be the only channel by which a Technical Establishment obtains operational experience. [Blackett, 1948]

So experience is taken from experiments, but the information to do experiments is taken from experience, not only from an historical data series that could have been registered for merely statistical purposes. This constitutes a hermeneutic circle or in Operational Research terms, a circle in the methodology. If we acknowledge the military background reflected in the literature, one can understand that Operational research as presented by Blackett gave a solution to classical problems in the strategic theory of military; for example "the difficulty of "accurate recognition" constitutes one of the most serious sources of friction in war, by making things appear entirely different from what one had expected. The senses make a more vivid impression on the mind than a systematic thought" [Clausewitz, 1976, 117]. The conflict between accurate recognition and systematic thought could be efficiently solved by a scientist of the profile of Blackett.

Again in section 5 entitled "Organisation and Personnel" uses the word experience, but with another meaning:

> A considerable fraction of the staff of an Operational Research Section should be of the very highest standing in science, and many of them should be drawn from those who have had experience at the Service Technical Establishments [Blackett, 1948].

And this is connected with one of the tasks of the Operational Research Section that was: to make possible a numerical estimate of the merits of a changeover from one device to another. For example, by continual investigation of the actual performance of existing weapons [Blackett, 1948]. This is connected with the numerical approach of scientificity, connecting section 6[13] with section 2 subsection i)[14].

Then in section 4 entitled "The variational method" he says:

13 One of the tasks of an operational research section is to make possible at least an approach to a numerical estimate of the merits of a change over from one device to another, by continual investigation of the actual performance of existing weapons, and by objective analysis of the likely performance of news ones.

14 The records of some war operation (e.g., air attacks on U-boats for the previous six months) are taken as the data. This is analyzed with quantitative possible and the results archived are explained in the scientific, sense, i.e., brought into numerical relation with the other operational facts and the know performance of the weapons used.

> The more common sense procedure is to abandon the attempt to construct from "first principles" a complete imaginary operation something like the real one under investigation, and to replace it by an attempt to find, both by experimental and by analytical methods, how a real operation would be altered if certain of the variables, e.g., the tactics employed or properties of the weapons used, were varied. [Blackett, 1948]

In this case experimental is referred to a testing of hypothesis based on "real" figures, that is, the observations already translated into numerical estimations, although the method for testing is not specified.

And finally in section 5.3 "Mixed methods"[15] in which experiment is linked with Geometry. Experiment in this case is the process to achieve a reliable result out of some proposed model. The testing of the model is what Blackett considers as experiment. As [Feyerabend, 1981] states:

> It is often said that the aim of science is the classification and the prediction of observational results, and that the distinctive characteristic of modern is mathematics and the experimental method. This is correct as a first approximation, but one has to add that experiments are used not only for discovering new facts, but also for revealing the detailed structure of facts already known [54].

This last part of Feyerabend's conception, is a further step of Operational Research in Blackett's work. Revealing the structure of already known facts goes in line of backing the views as in section 2

15 A desire derivative $\frac{dY}{dX_1}$ may often be obtained from another operational derivative $\left(\frac{dY}{dX_2}\right)_{obs}$ by using (a) a theoretical or (b) an experimental relationship between the assumed causally related increments dX_1 and dX_2 [Blackett, 1948]. Formally one can express the above method by the relation $\frac{dY}{dX_1} = \left(\frac{dY}{dX_2}\right)_{obs} \times \frac{\mathrm{d}X_2}{\mathrm{d}X_1}$ where $\frac{\mathrm{d}X_2}{\mathrm{d}X_1}$ is determined either theoretically or by special experiment [Blackett 1948].

subsection ii) of [Blackett, 1948] is explained[16]. Earlier Blackett puts into context the word experiment is in his [1935], where he explains the nature of cosmic rays and the variables that affect them. The text is in French, where there is neither phonetic difference nor graphic between the words experience and experiment.

> Les expériences avec les chambres d'ionisation. Les premières expériences consistaient dans la mesure du temps de décharge de la feuille d'or d'un électroscope, dans des conditions telles que toute possibilité de fuite électrique à travers les isolants fut exclue [Blackett, 1935,I-2].
>
> Experiments with the ionization chambers. The first experiments consisted in the measure of the discharge time of the gold sheet of an electroscope, under such conditions that all possibility of electric leakage through the insulating was excluded.

In words of Clausewitz, "Many intelligence reports in war are contradictory; even more are false, and most are uncertain. What one can reasonably ask of an officer is that he should posses a standard of judgement, which he can gain only from knowledge of men and affairs and from common sense. He should be guided by the laws of probability. These are difficult enough to apply when plans are drafted in an office, far from the sphere of action, the task becomes infinitely harder in the thick of fighting itself, with reports streaming in" [1976, 117]. The recognition of the use of common sense, the abstraction of war operations in a planning table, and the unreliableness of intelligence reports are convergence points of the rhetoric established between military and scientists.

An experiment for Blackett, in the cosmic radiation context, is the series of observations in which he is able to control the variables and measure them. In the experiment he could repeat the phenomena as many times as needed. For instance measuring the ionization effect on different altitudes 200 m. 250 m. and 270 m. [Blackett, 1935, I-6]. If

16 "The scientist in considering an operational problem, very often comes to the conclusion that the common sense view is the correct one. But he can often back the view by numerical proof, and thus give added confidence in the tactics employed". [Blackett, 1948]

this experiment was reproducible by others is another question different as to what Blackett was working on.

But what it is not an experiment for Blackett is the historiographic and perceptual approach called observation. In this sense observation runs in two senses: on the one hand, the Kuhnian approach for legitimating an established idea, and on the other, the necessity for further analysis of first time observations[17].

An experiment could be of many kinds. He was an expert doing this. He fits the concept and nominates denotes and connotes the term depending how the circumstances and the data of a problem are given.

In Operational Research the main case is: that the first approach was to observe and break down the data[18] and then to experiment not only to prove the validity of the hypothesis, but to experiment in the sense of reading and interpreting; two activities impossible to take apart in Blackett's world.

General Epistemology or the labyrinths of Mathematical methods and business development.

So far we have reviewed some crucial words taken as signs, the objects and their respective connotations in this brief semiotic analysis of the rhetorical resources of Operational Research during Blackett's times. These representations are fine for a mere linguistic analysis of the scientific texts, nevertheless to trigger these ideas in the context of history of science, i.e. to draw epistemic limits of the theories and deepen the explanation on the generation of knowledge it is necessary to connect it with some epistemological tools.

17 "When the actual figures become available through intelligence sources or at the end of the war, it will be interesting to see how far this estimate is in error". [Blackett, 1948, Appendix C]

18 In this sense, another convergence is the concept of intelligence and information. For Blackett is the main source of data and for Clausewitz is "every sort of information about the enemy" [1976, 117]

For example, in the area of Operational Research we cannot avoid the topic of maps, as maps are the representations of the real world by antonomasia. Moreover, our vocabulary on maps does change from our vocabulary referred to reality. We refer to objects in the map as the things themselves. We say (pointing to some area of the map) "This is Russell Square", as if the real Russell Square existed or "lived" in the map. Vocabulary referring to images works in a semiotic way let it be in science or just as common pictures as we have seen in the case of Operational Research.

The concept in colloquial vocabulary is that a map is a representation of the real life things. So the process seems to go from the "real world" to the symbols and drawings in a piece of paper.

The mathematical process sometimes works the other way around, building the model from the drawing to the "real object".

An analogy that illustrates this example well, is when we are trying to solve a labyrinth in a magazine or newspaper, a way to do it is to go backwards, i.e., to go from the centre of the labyrinth to the entrance, and avoid the hustle of trying paths that lead to dead ends. An important assumption in this case is that we know that there exists at least one path that connects the entrance with the centre, so we confidently just do it.

In the case of the construction of the map, we know, because we are dealing with a bijection, that there is an inverse function that invariably will lead us to the desired "mapping".

In Operational Research, the epistemology was more complex but similar, because they had more epistemic steps working in different levels, but the technique of solving the labyrinth backwards can be applied as an simile.

For Blackett to go from the physical dimension of the perceived "real world" to the first abstraction or the construction of the model he had to assume a certain behaviour and links between the model and the "real world", once done, the solution was an "easier" step. These assumptions were given by the experiments in physics and the scientific tradition, i.e., the world can be explained or "represented" by this model, let it be differential equations or distributions of probability.

Given this epistemic process, the reader might guess now, that semiotics in the internal-technical world of Operational Research run in the opposite direction: the signs lies in the abstract realm of the mathematical world, the connotation was given by the set of mathematical or abstract assumptions for that object, for example continuity, and the objects to which it referred were objects of reality.

The validation of the process is given by the relationship between the solution and the physical notion of the "real world". Once it was proven, the military or other applicants of the theory knew, as the solver of the labyrinth that there was a solution.

Solutions are probably the most important symbols (sign) that worked as tokens for both sides of the communication process (military and scientists); the crucial issue is that the discourse converged in such way that Blackett was a breakthrough in the discipline. Rhetorical analysis (texts) from this epistemological ground (construction of solutions) based on semiotics (words and their semantic context) can give a grasp of the success of Operational Research and its development to other areas of knowledge and of human activity, and why Blackett was considered by some the most important pioneer of this branch of mathematics.

Operational Research in this case worked backwards as if Blackett could see from above the labyrinth in which he was put and the discourse is so reassuring that now one can understand why a reader in those times could have felt confident on the content of the reports and recommendations even though he did not understand the substance, methods or details of the work.

The illusory distance decreased more and more, but in the end, the inverse discourse, as the Operational Research explanations, were there to produce a visibility, or to widen the vision. For example, in the appendix B of [Blackett, 1948], the treatment of derivatives is extrapolated to a real life situation[19]. This vision is a contribution that many other could not achieve (and maybe this was Blackett's secret of success), where the modernity of the world in terms of barriers of vision changed: The barrier was not the impossibility of seeing, but the impossibility of not seeing, and furthermore, the impossibility of foreseeing. Blackett's models adjust to this modern view, and manage the huge amount of visions and forecast situations.

Finally, in this epistemic process, we can identify three spheres connected with the semiotic analysis, in which Operational Research moves: a) historiographic, b) real, c) abstract.

These planes interact with each other, and are better illustrated in the following diagram:

Historiographic **Literary**	**Physity** **Real**			**Abstract** **Mathematical**
• Generates documents • Produces a style of coexistence scientific and non scientific • Generates a new sociology in the enterprises • Creates a new field of action for the scientist	• Generates a vision of the world, i.e. maps • Generates a visibility never had before • Generates a way of seeing the world			• Generate models • Create models • Implement applications • Generate solutions
Executive Technical	messages	messengers	interpret	Operational
New Organizations	• The idea of matter is produced by the abstractions and vision/imagination, in such as production of images	• Narrative identity in constant change • Imagination of becoming	• Meaning for words and generation of new ones • New semantics • Imagination of the past	• O.R. as a recognized discipline • However, it is not possible without the spaces producmed in the literary and real planes

The three spheres generate a literature, visions and a language, but they also generate entities, which generate again, in a second level, which we know more widely about Operational Research: new organizations, the models, the discipline, the documents and reports,

19 Since in each of these derivations it was verified that average value of the other variables was about constant, the four results represent in effect four partial derivatives. Making the reasonable assumption that these derivatives are causally significant, one can use them to calculate the relative value in saving shipping in 1941–42 of the four factors, numbers of escorts, and size of convoy, speed and air cover.

where we recognise the conflict between theory and practice in the application of mathematics to areas traditionally out of its influence. In this sense one of the most important issues concerning military strategy was friction, which was "the only concept that more or less corresponds to the factors that distinguish real war from war on paper" [Clausewitz, 1976, 119].

Chapter 3
The Paratextual side: Language and texts in the emergence of quantitative analysis.

Texts and Operational Research as a science.

In our epistemological quest of business development, we have to identify meaningful knowledge and its stages in order to have a clear differentiation of methodologies and the actual knowledgeable outcome. This distinction is failed by many authors who are still confusing science with technology/technique and have a main confusion of knowledge with information.

These issues are clearer in their actual exposition in documents, or more specifically in texts.

That is why in this chapter I analyze Blackett's position as an author and analyze his pioneering texts attempting to communicate results to high spheres of executives. I also consider the historiographic treatment of Blackett's documents as an epistemic problem in itself and make an epistemological classification of his texts, and finally I explain some of the nature of the operational research texts to achieve meaningful knowledge in the field.

Considering one of the approaches to science in the 20th century as an age of systems and statistics we have come to understand events in everyday life as interrelated phenomena occurring within abstract structures (see [Rau, 2001, 215]), nevertheless, we also find a difficult task to identify and distinguish who and in what circumstances the main characters of these scientific breakthroughs were.

In the case of Blackett and Operational Research, we find a complex structure with respect to this: we find that Blackett is a person, but also is an author and also is a character of a literary fiction (sometimes of his own fiction), and the problem lies in the fact that it is very difficult to distinguish them, in some occasions they overlap or interchange personalities in the same text, position or situation, for example: "Blackett is lecturing in 1962 about Operational Research and makes reference to his own book"; in this case we can firstly realise that there is a biological dimension, that is, we realise a fact, a fact that tells us that there existed a person named Patrick Blackett, the same that was lecturing in one of his areas of expertise, but in the course of the lecture we notice that he makes reference to another Patrick Blackett who is the author of a book in which it is narrated the "activities" of another Patrick Blackett who was one of the pioneers of Operational Research (in this case we have at least three Blacketts: Blackett the person, Blackett the author and Blackett the fictional character). To this scheme we can add several layers like another Patrick Blackett figurated by other authors like Lovell, who took him as a character of epic proportions, but this Lovell's Blackett is referred to the three other Blacketts, sometimes to the person, other to the author and so on, having an output at the end of another Blackett.

Some historians do not take into account this textual and paratextual activity, and perhaps confuse the author with the actual person; with that they contribute to dislocate historical accounts.

In the early years of Operational Research (England 1939–1945) important scientists like J.D. Bernal and P.M.S. Blackett gave a philosophical and organizational basis to the further development of what we know today as Operational Research; they produced a new approach to see/explain phenomena, which was hitherto within the domain of the military. Via ad hoc hypotheses and ad hoc approximations they created a tentative area of contact between 'military issues' and those parts of a new and powerful view (scientific-based view) which seemed to be capable of explaining them, not only issues concerning their nature and behaviour, but also the forecasts in the immediate future.

A crucial impact was the fact that this scientific approaches were aimed and even managed and applied by the non-mathematical audience, including the general public. In the cased of moving pictures, an interesting example is a dramatised situation, not far from historical reality, is to be found in the film *The small backroom* based in the 1945 book with the same title by Nigel Balchin, and with a production of 1949, only some years after the war, depicts in a way the activity of these scientists working with the military, as I said, just as in real life they might have done. In the case of the film we can see that the production or construction of scientific fictional characters was important in the sociological context; in other words, it was important to show to the general public a token of a scientific figure.

More than a technical basis or explicit methodology to develop the discipline, during these war years, it was launched a literature and a language, which gave an antecedent to the applicability of 'science' to other areas, like industry, transportation or agriculture; this was possible due to the extrapolation of the notions of 'science' (theories) in the process of perception of reality (application to real life) and language itself (the way of communicating it).

An interesting question arises here, and it is what was inherent to this literature that made it so powerful, to the point that its impact developed all the well known modern applications, studies, the construction of areas of expertise, and even the formation of societies of experts dedicated to research on different related problems but with the same sociological and methodological approach, and moreover, the institutionalisation of the name 'Operational Research' as part of the meaningful knowledge in the mathematical sciences. A first approach can be achieved via a rhetoric analysis:

Some of the facts that we generally know, and quoted by many other authors, like [Rau,2000] is that Operational Research has its origins during the Second World War in England, being P.M.S. Blackett one of the main characters involved, however the explanation of why Blackett was a main character in this story is sometimes flawed and left to tradition, in fact Blackett and his "circus" is referred by some, like [Ravindran 1976] as the beginner of the discipline, but none does not deepen more in the question.

One of the most important aspects that Blackett provided was that he created a notion of 'scientific' approach to general problems (or problems that hitherto were not considered of the scientific concern). Based on the limits of a writing activity itself we can say that Blackett founded a theory of writing as a saddle point—visual perspective—indispensable to establish an innovative idea in science[1], or if one prefers science as a fundamental reference of a literary language.

1 For more on this idea see [Hadamar, 1954].

One can trace back similar models, problems and applications of science in "normal life", however, the different aspect this time was that the problems did not become of a scientific area, but remained in the military scope. This implied that the merge scientist-military was achieved in the military environment and not the other way around. In other occasions problems passed to the influential sphere of science and became abstract problems proper of that science, scientists were working in similar problems but always remained scientists, i.e. never became part of a non-scientific organisation with respect to their work.

Blackett, in a way, proposed this saddle point and visual perspective concentrated in the problem of dealing with what was happening in the actual situations, in philosophical terms this is called the "becoming", which space corresponds to the set of facts, ideas and concepts that are actual and in the border of knowledge, or <expecting horizon>[2], i.e. the frontier that serves as field and object of study for science as conceived in the case of the emergence of Operational Research in Britain. This concentration in the actual situations might be the key to the posterior success of Blackett with respect to the creation of Operational Research.

In other words, one of the main characteristics of Blackett's textual analysis is the generation of 'vision' as such, a thing that was not usual in scientific or technical texts in those times.

Blackett wrote two documents during the war that were reprinted and quoted in some future works by him like in [Blackett, 1962], by his colleagues, v.g. [Waddington, 1973][3] and others like [Air Ministry, 1963]. The documents are entitled "Scientists at the operational level" and "A note on certain aspects of the operational research", where he describes, among other things, the nature, schedule, activities, problems, methods and tools of an Operational Research Section in the military service in the Second World War. These documents are considered the main source for the study of the history of Operational Research during these years.

2 The notion of 'Expecting Horizon' is taken from the prominent work of P. Ricoeur of 1984, *Temps et Recit*.

3 He became part of Blackett's team in 1942 and director of the operational research section in 1944.

Nevertheless, not many of the other documents authored by Blackett are mentioned in the literature of the history of Operational Research, for example other notes and working papers prepared by Blackett on the same subject, like reports that can be found in [Public Record Office, AIR15/297], minutes of meetings [Public Record Office, AIR 2/5352] or notes [Blackett Papers, Royal Society Archives, D64].

In any case, one must read "Scientists at the operational level" as well as all the other documents issued by Blackett and other pioneers in the Operational Research Sections of the Royal Air Force during the Second World War very cautiously in order to comprehend his underlying view and discover how the generation of knowledge in this field was achieved. During this time he was acting as a scientific advisor [Fortun 1993, 602], i.e. a person immersed "in and out" of the hitherto ordinary scientific environment and not just as a scientist who was called to solve some isolated problems or only someone to give some scientific opinion on some limited matters. This "double role" produced a transformation in that historiographic activity called science in the 20th century.

The present ideas are based on a cognitive ground, the axis of which is to be found in the textual form; therefore the analysis on the characteristics of these documents is present throughout all this historiographic view. Textual analysis is not just another area separated from the historiographical activity, for example [Bazerman, 1991], states that the rhetoric of science have emphasized the competitive struggle played by scientific text, or in other words, according to [Latour, 1987] referring to scientific publications as texts, they are seen as persuasive briefs for claims seeking communal validation as knowledge. Nonetheless, many modern commentators of science, as [Merton, 1973], make cooperation an essential component of scientific activity and communication, so the cultural and social components are not dismissed in the present textual analysis.

In the emergence of Operational Research, not only organizations, institutions and social factors affected the structure of the discipline, but also the ideology, knowledge, methodology, narrative resources and the construction of a sui generis epistemology as well.

A brief bibliographical account.

It is important to acknowledge the atmosphere in which the texts were written, the literary conditions in which Blackett was situated as scientific advisor were those of the institutional military commands in Britain, where likeliness, instead of logical validity worth more, and where the form of the text and the metaphors sometimes were more important than the theory behind it, and where also the bureaucratic network and political issues influenced in a meaningful way the work and production of the persons within. However, all of these factors are reflected in some way in the texts of Operational Research. In this sense it is obvious the importance of the study of the documents as texts as an axis of this work.

There are two primary historical problems to face when studying the documents: a) the epistemic problem of classification, and b) the historical classification of documents i.e. the classification made by Blackett, in which is reflected the kind of documents he chose to show and publish and the classification as an epistemic task, independent of any historical account, in which we can see what characteristics they utter to become the base of Operational Research without the support of journals like *Nature*, *Management Science* (which did not exist at that time) or any other specialized scientific group.

In the context of literary work, science or any scientific discovery as assumed in the 20th century must be in a written form and in these terms, writing is knowledge (see the work of Lyotard [1979]), and therefore an important factor of science is also writing.

Under this view, the concept of scientific legitimacy should be immediately taken into account. The etymology of the word legitimate comes from the latin 'Lex' that means 'Law', 'Legitimare' means according to the law or supported by the rules. In deciding what is *legitimate* or not, some of historians of science more dedicated to preserve the parameters of an institution, or even the same scientific community via some publications and preconceived ideas have linked this notion with that of a great theorem or great discovery behind a man or an institution, examples of this commonly made mistake can still be found in the literature.

The Legitimacy and Legacy of his work lie more in the hermeneutics and literary work built by the same Blackett some years after the war than in an external legitimation. To see this clearly, we must proceed with the first classification of texts in Blackett's opera.

Although there are many papers published by Blackett I will only focus on the most relevant for the Operational Research activities which are divided in a chronological way with respect to the war years, therefore we have three sorts of documents, the ones before, during and after the Second World War:

1. The previous ones, basically on cosmic radiation (as a matter of fact, he was researching on cosmic rays whilst he was appointed to serve on the Tizard Committee to begin with his work on Operational Research), where the influence in the way of treating problems is reflected directly in his posterior work specially in the Coastal Command and Admiralty. This bibliography is represented by two books "La radiation cosmique", and "Cosmic Rays: The Halley lecture"; some personal notes like the one depicting the use of statistical methods [Blackett Papers, Royal Society Archives, B24] where he uses the exponential method, the concept of curvature, and what he calls the check by used x' y' coordinates on the curved proton tracks; and other working papers, [Blackett papers, Royal Society Archives, B22] where he works on his thesis for the fellowship at King's College.
2. The documents issued during the war years. The documents we have in this period are of many kinds, but I would like to remark three: a) Reports made by the section or observations written by the persons in charge of sections in the Royal Air Force (RAF) commands. These writings are in the middle of a scientific and administrative paper. They were directed to military personnel and not to a colleague scientist or to the scientific community; b) Minutes of meetings held by the committees and sections, where a secretary reported the discussions about the problems and issues in the operational problems; and c) Notes on some specific topic, where he

describes the activities and organization of Operational Research Sections.

3. The posterior documents, regarding Blackett's activities in the Operational Research Sections during the Second World War. In this case we have some documents like "Operational Research, Recollections of Problems Studied, 1940-45", which appeared in "Brassey's Annual" in 1953 or "The Scope of Operational Research" published in the "Operational Research Journal" in 1950 which were gathered by the same Blackett in one book, "Studies of War" published in 1962.

In the following table it can be seen on the first column the kind of document issued, and in the first row the period in which such kind appears.

Period / **Type**	**Previous**	**During**	**Posteriors**
Letters	No	Yes, basically on organization of the sections.	No
Working Papers	No	No, except for Bernal	No
Theoretical Papers	Yes, in physics, cosmic radiation.	Yes, in form of reports/notes.	Yes, a couple of articles and conferences on the topic.
Minutes of Meetings	Yes, with Tizard committee on the defence of Britain.	Yes, in some committees and sections.	No
Autobiographical and Memoirs	No	No	Yes, as articles and including a book.

Reports	No	Yes, in a periodical basis and as notes to higher commands.	No
Notes	No	Yes, on request for some specific topic	No

In the previous period we only have two types of documents, of which the most important were the theoretical ones, from where he based his mathematical and "scientific" approach to operational research.

During the war, we have almost all kinds of documents, in Blackett's case there are not working papers, probably due to the secrecy related to the job, nevertheless in the case of Bernal one can find some working papers of that time. In a way the working papers are a source of legitimation, but in Blackett's case this notion grows from the posterior period.

In the posterior period to the war, we have the most important documents, which are the documents most quoted, well known and investigated by many: His memoirs and autobiographical accounts. This is meaningful in the sense of generating a narrative perspective, and even more a narrative identity; this means identify a position from which the subject that is writing is known. In the case of Blackett and Operational Research, as immersed in the military organization, the problem becomes a purely narrative one, therefore it is worthy to analyse it in this way. In this period is included the most important texts of the period during the war because these texts were reprinted several times in many publications including his autobiographic account "Studies of War" of 1962, where he narrates his experience as Operational Research leader in the military organisation years before.

In some way, the posterior documents are a kind of autobiography, in which a mixture of life, work and some kind of fiction is presented. The confusion between the general scientific knowledge and the stories related to it has to be analyzed as a historiographic account in the literature of science and not only as a social or linguistic phenomenon; this act of dismissal of some academics, for example, in Imperial College's Centre for History of Science, only limits the vision of historical research activities, as if the literature was not integral part of the culture of communities and an important sociological factor. It is important to notice that in papers like [Rider, 2000] the Blackett's work that appeared originally in 1941, which was reprinted in 1948, and in 1953 in different sources, and finally reprinted in 1962 by the same Blackett was referred by Rider as [1962], taking the same epistemological dimension of the one in 1941, in fact referring to the same ideas developed in 1941. It is clear that even if it is the same text reprinted, the historiographic treatment has to obey to certain epistemic principles, which some authors so not take into consideration. In other words, Rider does not make a distinction of important details like when, and from what position Blackett was saying "the same thing"; for Rider, it is the same to quote it from 1962 than from 1941. That is what happens when there is a lack of epistemic frames to base the historic work. And that is why also in this work; the same reference is carefully quoted and sometimes specified the reprints.

To analyse this historiographic account, and for an easier reading, textual analysis, especially for the posterior period where we find autobiographic documents, should be divided in two parts, as [Blumenberg, 1975] states: The "Lebenszeit" which is the life time, i.e. the natural biological dimension of the person who writes, and the "Weltzeit", which is the dimension of the world, i.e. the writable and buildable dimension in which the person who writes is immersed.

In Blackett's case these two dimensions are often confused, or ignored when talking about him; one can talk about Blackett in the service or Blackett in physics, but when life and work merge, all sorts of writings, be autobiographical or not, the narration of anecdotes or the heavy scientific work, must be taken into account in a special space generated uniquely by this work.

We cannot deny that Blackett was a political and public figure, a military officer and an academic for more than 30 years, but what we cannot either deny is that all this faces are reflected in some way in all his work. If he was a protégée of Henry Tizard or if his writings were controlled by the military, or if the famous quarrel Tizard-Lindemann was true or not, does not matter much because it does not affect the legitimacy of his work in the sense of historical activity and therefore literary activity.

Moreover, from a historiographic point of view, the writing activity goes further than the mere theory of writing; it reaches to the spanning of a space, i.e. a construction of a reality principle, in other words, the military and non-scientific personnel could see "reality" based on Blackett's explanations from another point of view. Blackett writes from the physical tradition quoting mathematical models used in the hitherto modern physics and using probability distributions, also widely used in physics, to explain precisely the physical dimension of the perceived real world (physity) of the <horizon d'attente>, where 'writing' and 'possible communication' constitutes the same thing.

In the posterior period, the theoretical and autobiographical papers including books, establish a concrete line in which the narrator (Blackett 1960) and the author (Blackett 1940) make a historiographic account of what was science, experiment and theory in the making of Operational Research.

In the previous period, the theoretical papers and the minutes of meetings are more in the scientific tone. The theoretical papers are totally into physics and the concepts, approaches and ideas on the problems treated in Operational Research are directly linked to them.

During the war period, the documents have many links, including the political and aesthetical factors. But it was in this period in which Blackett created these scientific-military reports readable or I would say translatable to the purely military spheres, that transcended as the basis of what we know now as Operational Research.

In the posterior period, we observe a paratextual activity referring to documents of the other periods.

The paratextuality in business planning.

Biography moves in two main directions, firstly the biological dimension of the man, concerning Blackett, this physical dimension means his achievements and presence in the political, military managerial and scientific life of Britain and the world, his Nobel Prize, his presidency in the Royal Society, his leadership in the physics department at Imperial College, i.e. the historical facts of the person and even more, the person as a fact and not as a historical product.

Secondly, the literary dimension, this looks at the character under some written account, i.e. the biographical review, the obituaries, memorial lectures, etc. This paper will be mainly concerned with the merge of these parts. The second direction will be divided in two parts:

On the one hand, Blackett's autobiographical work: the man, the writer, the physical person, and the author; this is of the utmost importance due to the coincidence in time in which quantitative methods had its boom, when the Operational Research Society was founded and the first journal of operational research issued its first number. In other words, the self-interpretation and the author, projected on some character built as an image of "Blackett-the-man" i.e. the person gave a whole approach and probably marked the coexistence between the scientists and their managerial counterparts in the business and war development activity.

On the other hand, the biographical account by friends and acquainted, like General Pile, Waddington, and probably the most important, [Lovell, 1976], which moves in different levels.

In the biographical approach, Lovell refers to Blackett-the-author, who was the founder of Operational Research, who is the first person in the autobiographical account. Blackett in his *Studies of War* functions as an author who refers to Blackett-the-historic-figure, who refers again to Blackett-the-person, the founder of Operational Research.

This division is clear because Blackett never says presumptuously "I am the founder of Operational Research" but via this literary resource it is the first thing one can infer, and in this role, the legitimation of his work is clearer.

He built a paratext not in the sense of clearing the chronology or establishing the order of publications, but to explain its philosophy, purposes and contributions, and more important, to establish a relationship among them.

One of the most interesting points about Operational Research is that it was not only Blackett who wrote his memoirs, but some of his

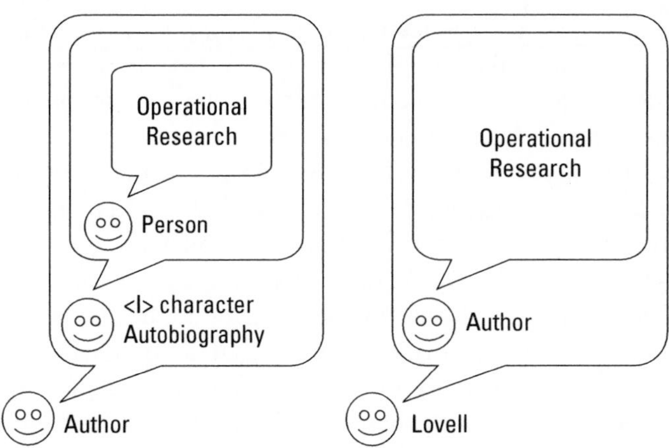

collaborators and colleagues did as well. We can propose some sociological reasons and effects of this, like the fact that including the scientific community, in special the mathematicians, accepted very well the generation of this accounts outside the specialized journals and purely scientific papers; but the most important issue to remark here is that this paratextual activity is also part of the creation and evolution of science, where the authors are also the first receptors of the theory, authors who have a well accepted proselytism that allows them to guide the first impacts of their work.

It was until later when the institutionalization of Operational Research, with specialists and journals and societies appeared.

The nature of the documents, a view on Military vs. Scientist.

Given the social and technological advancements during the Second World War, like the telecommunications, huge number of variables affecting directly war operations, inventions like radar or tele-directed more powerful bombs, one can say that the military had the necessity to look for advice, as they have been doing for a long time, the case of Galileo or LaPlace illustrate this point well, but this time the advice was needed at other levels, the level of strategy and more important the executive-political levels of the organisation.

As said in previous chapters, Blackett was more concerned in building a reality principle, more than showing a mathematical model to his military colleagues, this concern is reflected more in the political discourse than in the scientific or even strategic one (war strategic). It is not a scientific paper what he is aiming to do, but he is looking to make a reading of the real world with likeliness. In other words in the context of an epistemic discourse: he was constructing knowledge, out of the experiential world, i.e. based on some previous information, and not to merely doing the description of the reality independent of the interpretation of the scientist.

Consequently, there are three parts to observe in Blackett's documents: 1) writing about the <physity>[4] of the world via a scientific explanation, i.e. making a reading of the world, 2) the awareness of a translation activity which in a sense is a hermeneutical process performed by the military but induced by Blackett and 3) the scientific model; i.e. the reflection of the construction of ideas.

Summarising, we have the case of a scientific advisor who was actually invading the military's expertise area: the war. In this way, he had a varied job in the sphere of a writing activity, firstly in trying to explain, analyse, and evaluate different situations concerning operations of war and then dealing with the executive levels (the high military officers) which were "the experts" in the field [Rider, 1994], and secondly trying to convince them of an argument; these military officers at the same time were evaluating his opinion, this evaluation meant Blackett's survival in the bureaucratic system in which he worked. Blackett's immersion in the organisational system and other ideas linking operational research and other issues with the global

4 <Physity> refers to the physical stable system, which is the object of perception, it is where pure becoming takes place. It is the space in which subjectivisation begins but is not subject to it. For more of this term, see [Canaparo, 2000].

(national) political ideology is treated by [Jones, 1963] or [Kirby, 1997].

In other words, he had to show not only his point as in the scientific circles i.e. his truth in the sense of legitimation, but also "translate" his arguments in a systematic way so that the military officers (out of the scientific community) could match their values and previous ideas with this new view. In this sense, he might be considered genuinely as a writer because he is building a literary work, in the sense of the classical German philological tradition "literaturgeschichte"; and not only writing separated papers, in other words, trying to link several levels of discourse from the discontinuity formed between the scientific publishing and the military reporting worlds. Explained in this way, we can see how Blackett the author led to the basis of a new branch of science in the context of a modern world.

In fact, we could say that Blackett was the first to initiate a "paratext" in this field, i.e. to explain how the texts are conformed and what in the text must be read. In other words, Blackett did the reception of his own work; moreover he was the first receptor of his theory. For example in his [1950, 3] he highlights that he does not wish to discuss the history or its present achievements, but its methodology and organisation. The rhetoric analysis of this phrase tells us the way in which he positions himself in the field: when he says that he does not want to discuss about history (knowing that the history is himself) at the same time that he mentions the present achievements, he includes himself in the present achievements. Discussing on the methodology and organisation is the only left leaving us to imply that operational research is synonym of P. M. S. Blackett.

In this paper he continues by discussing if operational research was scientific or not, to what he says "yes by definition", having said totally the opposite years before when he left the service [Royal Society Archives, Blackett Papers, D120].

Then, having said that he did not want to discuss about the history of the discipline, he mentions the historical problem, that he solved, of the convoy size in the anti u-boat campaign. And finishes that Operational Research workers have to be in close relation with the executive authority, position of the utmost power in any organisation.

This did not happen in the scientific circles before Blackett, where there were standard values and knowledge pre-established to read and interpret some texts. In fact, authors were used to write for others, not for themselves, and many wrote for the future generations; this kind of authors, for example in philosophy Schopenhauer, expected the reception of his work after they died.

This "paratextual" activity might have given Blackett not only recognition of his work as the "creator" of operational research, but also a "registered trade mark" in the organizations and societies after the war, that is a product relying on the publicity and image management given by someone as the News Review of 29th May 1947, page 14, shows [Royal Society Archives, Blackett Papers, C68]. This note puts the photograph of Blackett side by side with probably the most prominent scientists of all times, Newton and Einstein, and with a big head saying "Newton, Einstein –and now Blackett" this note gives him the stature and authority to build a narrative from both inside and the outside.

Some others that help to improve his image were people like Tizard, who not only appointed him to participate in his committee during the war, and supported him in the spreading of his papers in the military circles, but also later in his activities at Imperial College, or Captain Wilder D. Barker, who gave a wide circulation of Blackett's first paper in the U.S. Naval circles [McCloskey, 1987].

In fact, many other authors, mostly in present times, have followed the same tactic. They become writers and historians of their own work, for example, Stephen Hawking, that even appears as a cartoon in "the Simpsons" labelled as "the most intelligent person in the world", then one can understand that his position can be almost unbeatable.

Under this scope, language is a crucial factor of change in the emergence and development of the discipline. Blackett's paratext, in the sense of validation criteria (legitimating), came out of the scientific explanations as creative mechanisms of propositions accepted as valid in the scientific circles, that culturally permeated to other social circles and organizations, which in a way, later became the objective reality, independent of the operational coherence of scientific work, i.e. reality then was referred to what was written. That is why in this work I sustain that Blackett created a reality principle by managing his textual activity.

The important epistemic step was to communicate to non-scientific personnel in the organization, as expressed by him, "One had to interact closely with the personnel at executive levels" [Blackett 1962], in Blackett's case the explanation was to tell them how to read the documents or tell them what to interpret about the concepts and ideas in the documents based on a correlation of the general intuition of reality; an example of this is the reprint of his two famous papers in his [1962] with the respective added comments. Other important details are 1) the fact that he never wanted to collaborate in any attempt to write the history of Operational Research, as proposed in the 1970s by Falconer, but instead answering that he had the intention of writing it by himself, [Blackett Papers, Royal Society Archives, D120] and 2) the defence of his work under some attacks:

In the following communication to Blackett posted in a letter from Sydney S. Dell, dated on 31st December 1958, Dell comments him about an article by Charles Hitch from the Rand Corporation California, U.S.A. (He was the Head of Economics Division, in those times Rand Corp. was in 1700 Main Street Santa Monica, California), who made certain aggressive comments on Blackett's work, but is not a direct as it seemed, Dell had to assume that Hitch was talking about Blackett, and most important, Blackett never answered in first person, he was always referring to "we", or 'it": ([ ]s are mine)

> First on his list [Hitch's list] of "really bad" errors in selecting criteria is your own work [Blackett's work] on the convoy problem during the war [although he does not mention either you or your own staff by name]. He contends that instead of choosing as your criterion of maximization of the flow of men and material across the Atlantic, you simply sought to maximize the ratio of U-Boat losses to losses of allied merchant ships. He says that no account was taken in your formulation of the problem of such considerations as the reduced operating efficiency of ships in large convoys and the danger of port congestion; indeed, it was an "intuitive restraint" rather than your analysis or the problem which prevented you from recommending the absorption of the whole merchant fleet and all destroyers into one single convoy. [Blackett papers, Royal Society Archives, D115].

And Blackett answered:

> What you say of his comments suggests a dangerously over abstract approach to these practical problems. We were too sensible to attempt to work out an optimum size of convoy. This would have meant knowing second partial differentials, which was out of the question. All we could evaluate roughly were a few first differentials. This allowed to predict that there would be gain by increasing the size of convoy. It would have been dangerous nonsense to predict how far this increase should go. For instance the enemies reaction was quite incalculable. [Blackett papers, Royal Society Archives, D115].

The use of the third person is also relevant in the textual analysis, as it indicates a frame in the sense of 'Sceneggiatura'[5], (see [Canaparo, 2000] and [Eco, 1979]), in which frame the subject, as a character, is situated.

The duality between symbols and ideas plays an important part in this process. Following Virilio, [1984], the speed of the technological advances surpassed the normal human activity; reality was no more but long lost ideas that could only be represented by images, icons or symbols. In other times reality of the daily life was more tangible, this means that a soldier could know how to fix a cannon or knew how to use the tools in a battlefield. Even more, his knowledge about how the war was going was as accurate as his perception was.

To the point of his (physical) perception the soldier could know if his companion was wounded in some difficult battle in the mud or under the rain, he did not need of weather statistics or an enormous list of wounded reflected in numbers, every soldier could feel his "reality". A General could know (in the battlefield) if the war was going in one direction or another if the number of killed or injured was for this or that reason and support his argument by showing physically the facts and claiming to have been there personally.

The way in which scientists worked (in mathematical models) was by trying to give a general solution to a problem in a higher level of abstraction, and that is more a qualitative method than achieving a numerical result from a particular problem, as said by many, for

5 This term refers to a surrounding or to a scene placed like in a theatre play, in which the character is situated, this scene is constructed by the same speech and makes reference to the position of the narrative identity.

example, [Kittle, 1947] who proposes the following definition of Operational Research: "Operations Research is a scientific method for providing executive departments with a quantitative basis for decisions. Its object is, by the analysis of past operations, to find means of improving the execution of future operations" and continues with the historical account of the discipline. After having analysed and research definitions from the early 1930s to the end of the Second Word War, I have not found any definition of Operational Research even similar. None of the words he uses, like "scientific method", "executive departments", or "quantitative basis" is used by the commands even at the end of the war.

These quantitative methods are proposed by Blackett only to back up arguments as the opinion of an expert. The only tricky thing was that Blackett had to put this scientific analysis also in the terms of reality as exposed in the above paragraph (page 19).

In operational research numerical data and quantitative analysis were important, but the translation of qualitative properties such as symbols, concepts, meanings, representations, and in general the semantics of these different levels were central in the development of the discipline, because it was the vehicle to make it understandable and applicable for the executive levels of the organization who were not scientists.

The language has to deal with some rules of translation, not only paraphrasing in the same terms of an internal science, as Putnam puts it in his [1975], "the rules of translation are so-called definitions which appear in formalized systems. They are best viewed not as adjuncts to one language but as correlations between two languages, the one part of the other".

Translations, went one step beyond with Blackett, they gave meaning more than simply naming the concepts, they offered ideas and properties to the objects treated in all the operational research experience in the coexistence of scientists and military.

The way of doing it had its key point in the documents themselves. In the above-mentioned documents (military reports), one can find not only topics concerning war activities directly, but also a lot of scientific ideas as well as other kinds of concepts (like organizational or administrative).

It is worthy to note that the method is on the construction of criteria. What Blackett does is to describe a method of reading, as said above, a "paratext", but not the methodology itself, which was given for granted and called 'scientific'. This is another misreading of Blackett's texts, for example [Rider, 1994] where he says that Blackett, in his two reports drafted in 1941 talked about the methodology of Operational Research.

However, as a person not familiar with operational research terminology, like notation and concepts (scientific-mathematical), one can learn by reading Blackett's reports and notes about mathematics, probability, modelling, differential equations and war operations, and more important, one can be aware of the intention of the reports with respect to the activities: "an Operational Research Section can act usefully by interpreting…" or "the scientist can often back the view by numerical proof" [Blackett, 1948, 28]; Blackett's purpose was not to model, or forecast, (actually these are by-products in the epistemic process), but to interpret and support numerically an opinion. The paratextual frame explain then how ideas like the ones exposed in [Kirby, 1997, 558] about the elements of quantitative analysis to understanding the source of victory and defeat in the military operations are widely believed as linked with the emergence of Operational Research and the work of Blackett.

One cannot find in Blackett's documents, an explicit precise methodology. The documents are not trying to explain more than what the process leads to. This means that they do not discover anything new from reality, except what in a natural way comes from the techniques used: The results. In fact the objective of operational research was in Blackett's view more practical (note that it is not the objective of operational research, but of the operations analysis):

> The objective of operations analysis is to identify the significant variables in the operation of any system, evaluate their significance with the aid of observational data supplemented by theory, and draw conclusions as to what variations will lead most effectively to the desired result [Blackett Papers, Royal Society Archives, D113].

If on the one hand the documents do not refer to the methodology itself, i.e. they are not proposing a way to study or an explanation of the problems, but rather are reporting the results derived from some "hidden" methodology not specified in the documents (although they are attempting to report on the methods and operations of the section: "it may be of some value to discuss certain methods of approach", [Blackett, 1948, 29]); on the other, they have the power of adding concepts from a wide variety of fields (not only mathematics), for example psychology or biology and further more a human factor, without losing his part of a "rigid" scientific paper[6].

For example, the documents do not explain how to "decipher" the tables of information but they only say things like: "This process is one of extrapolation from known data" [Blackett, 1948, 32], but they do not mention the way of extrapolating (I can figure out at least 3) or "The first step in the analysis is to break down the statistics… in such way as to give their variation with the main variables" [Blackett, 1948, 35], the question is: How? What main variables?, how do I choose them?, In what way should I have the information and treat the data tables to do so?, What is a variation? Etc. Blackett does not expose a methodology, so in order to know it, one has to go analyzing report by report and extrapolating with the existent theoretical background (i.e. cosmic rays, models in physics and mathematics, dynamical systems, topology, etc) to find out the way he managed to achieve a result, and further more, to be confident of it; this is why the epistemological view is so important in this case, not only to discover the process of achieving knowledge but to "unpack" their methods[7][8].

The documents illustrated very well the logistics, organization, and quantification of variables in war operations, and to do so, they had to change the form of the military reports hitherto issued, but in no way they give a precise methodology.

6 For an example of scientific/mathematical paper of that time, see [Birkhoff, 1927].

7 In this case we will have to deal with the logic of empirical methods and more epistemological and mathematical arguments (like linear algebra).

8 Nevertheless, methods like differential equations and dynamical systems were widely used to solve, describe and explain phenomena in nature; for instance, the work of Blackett in Cosmic Rays.

In fact, as discussed before, operational research as conceptualized by Blackett was concerned with the organizations and the quantitative analysis but not as a main goal, but always taking into account other important factors like theoretical or mathematical.

There are two points to note in this sense, first the way in which they are synthesized and second the new layout for documents; this meant a change in the form and style of the documents which was a discontinuity in the scientific literature and led to a closer relationship with the non-scientific circles, for example in the industry, economy, government, etc.

As mentioned sooner, the underlying purpose of the documents was not to illustrate on the methods of doing the analysis (there is no document explaining the methods or exposing algorithms on how to treat the military problems), because the problems were taken as 'situations' to be explained (they did not need algorithms to solve anything), but the documents generate an epistemic construction, independent of the well recognized models, in other words, they built up a language.

In general Blackett was not asked to solve a standard problem, but to explain a situation; this does not lead to a scientific theory, but to a narrative identity, that generates a writing construction, that tries to interpret reality; in fact, in many cases the same scientists did not know anything about what the problem was about. Again, in this sense cosmic radiation studies have a lot of parallel concepts with operational research: Blackett had first to deduce, out of nothing, what possible variables could affect this phenomenon and these could only been inferred from photographs or some data collected by the observer, but definitely this is not a problem-solving situation.

In this process the Operational Research Sections could be responsible for creating the problems, in other terms, to generate their own narrative space based on a series of physical images and create their own identity/subject:

> This group [the operational research group] was not the first group of civilian scientists studying operations, but it was certainly one of the first groups to be given both the facilities for the study of a wide range of operational problems, the freedom to seek out these problems on their own initiative, and sufficiently close personal contact with the service operational staffs to enable them to do

> this [Operational Research, Recollections of Problems Studied, 1940-45, reprinted from "Brassey's Annual", 1953, pp. 88-106, Blackett Papers, Royal Society Archives, D59].

That is why it is shown how analysing his rhetorical resources Blackett built a paratextual platform from where the general speech of Operational Research his work was inserted in the literature of the discipline as the main foundation. This last point gives an explanation of why authors like [Waddington, 1974], [Dando, 1978] and more recently [Kirby, 2003] argue without rhetoric specific analysis, that Blackett had an anti-scientific attitude, by opposing to the mathematisation of operational research. This argument, as we saw, is totally out of context given the identification of the narrative identity, vindicating Blackett's work in this sense and at the same time allowing to appreciate the epistemic projection of the pioneering work.

Chapter 4
R. A. Fisher: experiments and mathematics, a multidisciplinary view.

Introduction.

The work of R. A. Fisher covers a wide range of topics. He developed methods on statistics and experimental sciences and the use of his methods have been influential in many areas of knowledge, in fact, they can be applied to biology, genetics, agriculture, industrial production, pension plans, and medical trials.

In 1925–1935, probably the most productive time of his life as book publishing is concern, he published the 3 main books about statistical treatment of information; the present work is mainly centred on that period.

To discuss his wide area of work with so many varieties and components: social, scientific, experimental, theoretical, etc. with these factors, in addition, closely interrelated, it is hard to grasp it from one point of view. Nevertheless, if considered from the knowledge viewpoint, one can see it under a general epistemic view. This <general epistemology> describes the process in which experimental science and applied mathematics converge.

In Fisher's work experiments and theoretical issues were first formalised in his book *The Design of Experiments*. This book will be analysed in detail in this Chapter. It is worthy to note that, cases in which these two areas (in this case experiments and mathematics) mix, they have similar epistemological frames which are those involving the construction of a narrative which is always subjected to hermeneutical study and where representation as a basic epistemic step is implicated. However, one comes from theoretical notions and the other from empirical activity. In this Chapter I will try to show how these two areas, apparently disjoint, have parallel epistemic processes.

It is in the formalisation, systematisation and ordering of the empirical observations, where abstract concepts of mathematics seen as models of the "real world" meet, where the cognitive act (as part of a general epistemology) gives birth to a new vision and approach to what we call applied science by generating a space in which measurement as a formalised theory creates a expecting horizon on which other cognitive acts will take place. In this sense the cognitive act is a chain of iterations of the same epistemic process. That is why measurement, in this sense, does not take place in the "real world" or with any objects of the physical existence, but in the same formalised theory that manipulates the measures. The only link with the "real world" takes place at the time we decide the units of measurement, which is a technological act, cognition is given after the dealing with the world has been achieved. An example of cognition is the act of proving (which takes place in the space generated by the cognitive act itself.

This <general epistemology> is one of the axes of the present work as it is a common factor in the "scientific breakthroughs" of Fisher.

This work covers three major breakthroughs achieved by Fisher, all related to applied mathematics:

1. The correlation coefficient. This measure is concerned with the relation between variables in experiments. In the present work this will be treated under a geometric approach, although typical analysis have considered it as an algebraic problem in the field of statistics.
2. Analysis of Variance. The central topic is, in this case, the calculation of the variations on some observed data, which is seen as information. In fact the notion of information, distinguished from the notion of observation or data, is crucial to understand some of the main concepts in statistics, such as this. Fisher concentrated, not only on the probability of events but also on the probability of errors and other deviations associated with such events, and it is under this view that the analysis of variance acquires its utmost importance.

In this point I will also consider from a geometrical point of view his contributions in the analysis of variance as a technique for the separation of sources of variation in the sets of observations.

3. Design of experiments. In Fisher's view, the basis for statistical analysis is experimental evidence. Due to his position in the Rothamsted Laboratory at that time, he had at hand the resources to supervise and modify the experiments done in the field. This happened, to the point that the two concepts mentioned above depended on the experiments from where the data came. His book The Design of Experiments, published in 1935, is mainly concerned with the formalisation of methods to observe and to gather data, establishing the object, objective and methodology of experimentation as well. This is why, for the purposes of the present work, the design of experiments will be considered to an epistemology in itself.

Throughout this Chapter I will discuss the relationship between science, technology and technique. This is because Fisher's work revolves around concepts on the border of the theory and practice of statistical methods (science and technique), but also with their presentation and explanation. I also will discuss the work that covers the application of these concepts within the frame of observation and experimentation.

The three points above have interconnections at different levels, so it might happen that some concepts and areas may overlap or that they are cross-referenced. In order to explain them in a clearer way, these points are analysed in the following six directions:

1. Epistemic. Analysing the processes of construction of knowledge as well as the means and ends of concepts and generation of notions.
2. Ontological. Concerning the connections with reality and the level of existence of inputs, outputs and algorithms in relation with their applicability in "real life".

3. Hermeneutical. Analysing the role of the scientist as interpreter of texts, facts and translator of concepts.
4. Semiotic. On the generation of special signs and their meanings in each iterative process.
5. Language and Linguistics. Analysing elements of language involved in the communication of ideas.
6. Literature. Analysing literary style and writing activities of scientists in their pursuit of transcending from theoretical mathematics to applied fields. Some of them belong to the humanities or to social studies.

My arguments will revolve around language and literature, which I consider the vehicle of knowledge, but will draw from the epistemic and hermeneutical areas, based in a constructivist theoretical outlook.

A brief discussion on the object and aims of probability.

It is important to reflect on the view of probability and discuss on the objectives and objects of it in order to understand the position of Fisher's work, and explain the deductions and aims of his work.

Fisher had a strong position against the *Inverse Probability Principle*; this view was present since the first years of his career. This is important to mention because it could have been a great influence to develop his later work, on the other applied areas.

In the fourth edition of *Statistical Methods for Research Workers* published in 1932 he adds in the introductory chapter a section entitled "historical note", where he makes a brief account of two methods of doing statistics, on the one hand, the method based on the *Inverse Probability* and the other based on empirical methods on observations.

He also states assertively, in the introduction of the fourth edition, that he is against the Bayesian point of view and the principle of indifference.

> For many years, extending over a century and a half, attempts were made to extend the domain of the idea of probability to the deduction of inferences

> respecting populations from the assumptions (or observations) respecting samples. Such inferences are usually distinguished under the heading of Inverse Probability, and have at times gained wide acceptance. This is not the place to enter into subtleties of a prolonged controversy; it will be sufficient in this general outline of the scope of Statistical Science to reaffirm my personal conviction, which I have sustained elsewhere, that the theory of inverse probability is founded upon an error, and must be wholly rejected. [Fisher, 1932, 10]

Mathematics is a discipline that is characterized by the use of a specialised language. This language sometimes has no direct translation or equivalence in the common language, for example some concepts in spectral theory or in abstract algebra are used only in a mathematical context. Sometimes the language of probability "lives" in that area; for example in calculating the probability of choosing among the Natural Numbers a prime. In this case the classical view of frequentists does not apply because the number of possible cases is aleph-zero, the same happens when trying to calculate the probability of choosing a point in a circle; there are as many possible cases as there are points in the circle.

For Fisher Probability was applied to more "real" and practical cases. Although he takes the language of the abstract inner mathematics, he extrapolates it to cases in the areas of agriculture or genetics. It is important to note that if one applies the concepts of probability to a "real problem" one is assuming the properties of the mathematical entities to the real dimension of the world (and not the other way around); otherwise the model would not explain but theoretical thoughts. For example if we assume something to be describe as a "Normally Distributed" we extrapolate the properties of the Normal Distribution which is continuous, integrable, differentiable, bounded, etc. to the objective case, moreover we assume that secondary properties like those of its mean and variance, such as the linearity, are valid. The two cases are mixed, for example, analysing the case of continuity, on one side (real case) the representative points are very numerous and very close together, like the molecules of matter in some atomic model, and on the other side (mathematical theory) the representative points are replaced by a continuous matter; the distance from "reality" is large.

Before the 20th century, some mathematical theories that explain natural phenomena were based in abstract and a-priori work, for example, on the tools of calculus and differential equations which are deterministic models in the sense that the functions are variable dependant. However, mainly in the 20th century, probability and the concept of chance acquired a very particular importance, particularly in physics, for example, in quantum mechanics, radioactivity and cosmic rays.[1]

Fisher's work is focused on probability and statistics and their applications to non-mathematical areas[2]. Fisher considers probability in the field of statistics:

> The deduction of inferences respecting samples, from assumptions respecting the populations from which they are drawn, shows us the position in Statistics of the **Theory of Probability**. [Fisher, 1925, 9][3]

The concept and the essence of probability are crucial in his work to define his view of statistics. In this way one can understand how he developed methods such as the design of experiments and concepts like the correlation coefficient.

These ideas are contained in a statistical frame, which is the general aim of his research. Most of the time he deals with data coming from samples and, as [Segal, 2003] says, there is a conversion between the notion of data of a sample and the notion of information, which in the case of Fisher is the core to develop models in mathematics.

The value of a probability-based approach is the reduction of the amount of required knowledge. To illustrate this, we can look at an alternative: The deterministic approach. In contrast to a probability-based approach, a deterministic approach requires full knowledge of all

1 Nevertheless, methods like differential equations and dynamical systems were widely used to solve, describe and explain phenomena in nature; for instance, the work of Blackett in Cosmic Rays.

2 Although these areas include physics, the arguments will revolve around other natural science like biology, genetics, medicine, and applications to areas canonically classified as humanities, like economics and management.

3 Bold is Fisher's.

variables and, from this point of view, it is argued that any outcome can be calculated. For example, the initial position of the dice, the friction with the air, the angular momentum, the force with which the dice is thrown, and the other factors involved in the phenomenon, and from this, one is able to deduce in advance the exact outcome of the event. Unfortunately, the control and knowledge of all the factors involved in the making of an event is impossible; sometimes because the variables are too many to even calculate a preliminary result, as in the three body problem, or because some of the variables are not quantifiable, like the morale of a soldier in war, or the psychological state of some observer involved in an experiment. Thus, the notion of probability has become an important tool in the solution of complex problems like these.

Within probability-based approaches, and in the early 20th century, two contrasting strands were evident. One of these, nominated by some "Bayesian", is founded in an a-priori assumption: The principle of Inverse Probability. This considers the probability of an event equally distributed with other possibilities of outcome. The other is the frequency distribution, based on actual experience or observation of the phenomena. For example, a Bayesian would assume that the probability of getting a "5" out of throwing a dice is $\frac{1}{6}$ (equally distributed to all possibilities, i.e. one for each face of the dice), or $\frac{1}{2}$ to the probability of getting a head in a toss. A "frequentist" would argue that there is no evidence in which to support that the probabilities of outcome of some event are equally distributed; therefore he/she needs experience of that event. In this case, the observer turns into an experimenter and begins to fulfill records with outcomes of some event, for example tossing a coin, and out of that bulk data, he calculates an initial probability, which could be $\neq \frac{1}{2}$.

And although the conclusion of an experiment might be that the probability of getting a "head" in a toss is $\frac{1}{2}$, the difference between these two approaches to probability lies in its epistemology and it is not purely nominative, and these views are the basis for analysis of meaningful knowledge, such as that done by experimental workers, generated by scientists in the early and mid 20th century.

The argument concerning the theoretical foundations and analysis of probabilities becomes even more complex. On the one hand, in the Bayesian approach, due to lack of information, the arbitrary a-priori assumption on the probability of some event could be wrong and lead to biased and erroneous results. On the other hand, the frequentist has to deal with errors and inaccuracies in the observations: that is in the observer (who is human), the instruments (built and used by humans, and therefore susceptible to mistakes) used to observe, the number of observations and the treatment of data.

In an environment in which observations and empirical experience were tools to achieve theoretical knowledge, Fisher's work was of the utmost importance, especially for those who were interested in the concept of <objective truth>. In this sense Fisher is another scientist that links the term (<objective truth>) with the term <scientific truth> referring to the knowledge that was expressible in terms of probability and related (not necessarily confirmed) in experimentation.

In the following diagram one can observe the object of probability according to two different views and objects surrounding an event. Fisher is aware that probability is more useful in an applied science when calculating also the probability of the circumstances surrounding an event:

There was also another circumstance which stood in the way of the first investigators, namely, the not having considered, or, at least, not having discovered the method of reasoning from the happening of an event to the probability of one or another cause [De Morgan, Cabinet Cyclopaedia, quoted from Fisher, 1935, 5]

To understand the points of view and interest of Fisher and other scientists concerned with the application of science in the 20th century, one should answer the question of why probability is important for the physical sciences and humanities. A first consideration is the powerful property of predictability that the calculus of probabilities gives. A second consideration is the inferences and generalizations that can be reached through the theory of probability in areas like physics or those more in contact with human activities, such as economics, war operations, agriculture and development.

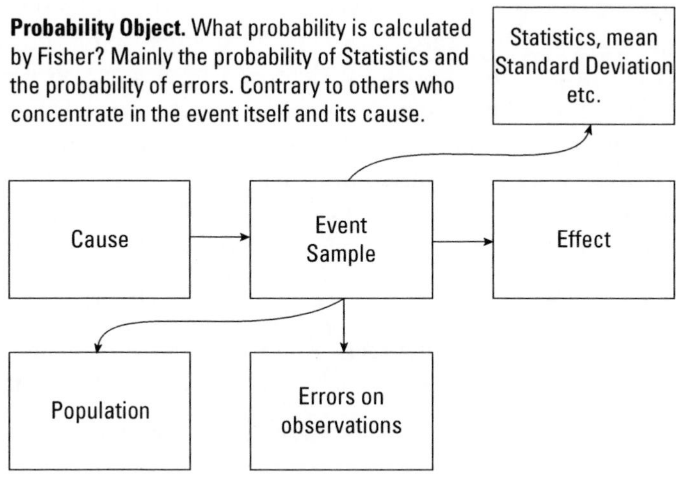

Furthermore, it is certain that the calculus of probabilities is a useful tool to decrease the level of ignorance we have of the problems. If we assume that the universe behaves analogically to a dynamical system, i.e. with a law of change and iterations of this law, and if we know the initial state of the system and the function that generates the subsequent states, the problem becomes only an algebraic or purely mathematical problem. But in many problems, the ignorance of some of the variables, for example, the initial conditions of the problem, or the law of transformation, can make our conclusions and statements about the phenomena erroneous. Nevertheless, in some cases, like in the kinetic theory of gases, the degree of ignorance does not allow us to conclude anything with certainty: within this theory it is assumed that the molecules follow rectilinear trajectories and behave according to laws of impact and elastic bodies, but we do not know the initial conditions of the velocity, momentum, position, not even the present velocities of the particles and therefore do not have enough information to determine the general state of the system. The calculus of probabilities, via the mean phenomena resulted from the combination of velocities, might tell us this state and the behaviour of the system with some degree of accuracy.

In some cases when a scientist requires the process of induction to conclude on some area, probabilities may be of great help to do so. But the calculus of probabilities can be dubious and arbitrary as well, not only in the view of the frequentist concerning the management and performance of experimentation, but also in the Bayesian view: For example, if we want to calculate the probability of getting at least a "5" from throwing two dice, we calculate the number of possible cases, which are 36 and the number of favourable, which are 11, thus the probability of getting at least a "5" is $\frac{11}{36}$, however another view of the problem is calculating the number of combinations of two dice form, that is 21, and the one that contain a "5" are 6, therefore the probability is $\frac{6}{21}$. The two probabilities are correct in their own view, but not only that, the epistemic argument turns more complicated when we assume that the events are equally probable. This can turn into a circular definition, as we define the probable by the probable, or calculate the probable assuming the probable[4].

Having agreed in a convention the rest is just given following some mathematical rules and algebraic manipulation of expressions. The problem comes when the scientist wants to apply this result to real life, then he has to argue his considerations, conventions and methods, in order to legitimise the result, i.e. the results in paper have to coincide with the real world. Probability has a high practical value; it leads to courses of action and decreases the argument of ignorance on matters that require accomplishment.

After this general scope of probability and the opinions of Fisher, one can see that the discussion of the views, methods and management of data are cause of considerable dissent, so one of the most important epistemic steps is the communication of these ideas. In the case of the work of Fisher the discussion comes from the first observation as in his book Design of Experiments he basically teaches the reader how to <observe>, which might sound naïve and irrelevant, but as seen above,

4 See [Poincaré, 1903]

it is extremely important to sustain a solid theory when concluding from probabilities.

Moreover to make a final comparison on this area, on the one hand, some other scientists contemporary of Fisher try to calculate the probability of the occurrence of an event, meanwhile Fisher is trying to calculate the probability of the cause of that event; and that is the essence of the retrospective method of inference.

On the other hand, both are trying to forecast in a way the probability of the effects of that event, in some cases already given (like in the case of Inverse Probability). So experiments and experience are in a similar level of observations-data, the difference is that experience is denoted for the accumulation of bulk data, and experiment to the way in which the recollection of data was done. In this sense we can differentiate as well the concept of information and of data from samples.

Brief biography of R.A. Fisher.

Ronald Aylmer Fisher was born on 17 February 1890 in East Finchley. From a young age, he suffered from extreme myopia, and even in his school days his eyesight was very poor, so he was forbidden to work by electric light, developing the ability to solve mathematical problems without any visual aid and an extraordinary geometrical sense.

He attended Gonville and Caius College Cambridge in 1909, and became a Wrangler in 1912. After graduating he spent a further year at Cambridge, studying statistical mechanics and quantum theory under James Jeans and the theory of errors under F. J. M. Stratton.

Although Fisher specialized in mathematics in the years of his education, from an early age he also had a strong interest in biology. Later, he became keenly interested in evolutionary and genetic problems.

On leaving Cambridge he took up statistical work in the office of the Mercantile and General Investment Company. From 1915 to 1919 he worked teaching mathematics and physics at various public schools.

In 1912, he published a work concerning the fitting of frequency curves. In 1918 he published a monumental study on the correlation between relatives on the assumption of Mendelian inheritance. This was first submitted to the Royal Society but on the recommendations of the referees was withdrawn and was subsequently published by the Royal Society of Edinburgh, partly at the author's expense.

This work led to simultaneous offers in 1919 of the post of chief statistician under Karl Pearson, at the Galton Laboratory, and the post of statistician at Rothamsted Experimental Station, under Sir John Russell. He chose Rothamsted thinking that he would have more opportunities for independent research there and also to be able to pursue genetic studies more actively. He was right, Rothamsted provided an exceptionally free atmosphere for the pursuit of research, and brought him into close contact with biological research workers of very varied disciplines and attainments.

Fisher's appreciation of, and readiness to discuss, the practical needs and difficulties of other workers in the laboratory soon began to bear fruit. He developed an original idea of statistics, a notion of experiment and a conception of the application of mathematics (via statistics) to other sciences. This point of view might be compared with other important achievements such as Blackett's, or Kantorovich's concerning its social role and its intensity of evolution on other branches of knowledge and activities of people at the time.

What lies behind Fisher's work is the quest for the discovery, not only of the causes and effects of events, but also of the interruption of the natural form cause-effect. These anomalies were the main object of study for Fisher. Probabilities and statistics (as the study of populations) are the essence of the treatment of errors. His contributions to statistics are more concerned with the theory of errors, which in his opinion was one of the oldest and most fruitful lines of statistical investigation. [Fisher, 1925, 3]

His ideas concerning the management of variables are based in a process of dealing with the deterministic (in the sense that there exists a causation) and the chaotic (in the sense that there is also a probability of occurrence); thus he explains how the reduction of variables in the process of empirical methods are achievable via the establishment of a correlation between them:

> If we know that a phenomenon A is not itself influential in determining certain other phenomena B, C, D, …, but on the contrary is probably directly influenced by them, then the calculation of the partial correlations A with B, C, D, … in each case eliminating the remaining values, will form a most valuable analysis of the causation of A. [Fisher, 1925, 153]

While at Rothamsted not only did he recast the whole theoretical basis of mathematical statistics, but he also developed the modern techniques of the design and analysis of experiments. He was prolific in devising methods to deal with the many and varied problems with which he was confronted by research workers at Rothamsted and elsewhere.

The book *Statistical Methods for Research Workers* which, as we have said, appeared in 1925 (at a time when he was already Chief Statistician at Rothamsted) published by the prestigious publishing house of Oliver and Boyd. According to the editor in the preface, the book was intended to be used as a text-book. It was accessible to biologists, who were not slow to take advantage of the new methods presented in it. Four yeas later he was elected Fellow of the Royal Society and he became known to a much wider circle of research workers.

While at Rothamsted Fisher pursued his studies on genetics and evolution, and undertook a series of breeding experiments on mice, snails and poultry; from the last he confirmed his theory of evolution of dominance. His book The genetical theory of natural selection" was published in 1930; in it he attempted a reconciliation of Darwinian ideas on natural selection with Mendelian theory. He also developed in it his theories on the dysgenic effects on human ability to be the parallel advancement in the social scale.

In 1933 the importance attributed to his genetical work led to his appointment at University College London, as Galton Professor, in succession of Karl Pearson. However, as such he did presided over the whole of Karl Pearson's former Department. Karl Pearson's son, Egon S. Pearson, was put in charge of the statistical section, which was renamed the Department of Applied Statistics.

The Galton laboratory offered opportunities for the experimental breeding of animals which had not been available at Rothamsted. In association with G. L. Taylor and R. R. Race, he developed the study of genetical aspects of blood groups, and in particular unravelled the complexities of the Rhesus system, which is responsible for erythroblastosis foetalis.

He also took over from Karl Pearson the editorship of *The Annals of Eugenics*. This journal had originally been founded for the publication of papers on eugenics and human genetics, so that *Biometrika* could be left free for papers on exclusively devoted to statistical methodology; under Fisher it rapidly became a journal of importance in statistics.

In 1935 The design of experiments was published; this was the first book explicitly to be devoted to this subject. In 1938 he and Yates produced "statistical tables for biological, agricultural and medical research" and in 1939, at the outbreak of war, University College was evacuated and he returned to Rothamsted, where Sir John Russell found accommodation for him and for his department. However, in 1943 he accepted the Arthur Balfour Chair of Genetics at Cambridge, in succession to R. C. Punnett; Fisher held this chair until his retirement in 1957.

During his later years Fisher was the recipient of many honours. He was awarded the Royal Medal of Royal Society in 1938, the Darwin Medal in 1948, and the Copley Medal, the highest award of the Society, in 1955; he was knighted in 1952. He was made an honorary member of the American Academy of Science, a foreign associate of the National Academy of Science of the United States of America, a foreign member of the Royal Swedish Academy of Science and the Royal Danish Academy of Sciences and Letters and a member of the Pontifical Academy of Science. He also received recognition from several foreign universities. He died in 1962 leaving a large body of work which can be divided into three main parts: statistical contributions, the design of experiments and Genetics.

His respect for tradition, and his conviction that all men are not equal, inclined him politically towards conservatism, and made him an outspoken and lasting opponent of Marxism, a movement which, at the time, attracted the attention of some British academics.

The interpretation of Fisher as an author is an important perspective of this work, as he is trying to project himself into the world of the social sciences. Even if efforts had already been made in the previous century, this was still a largely unexplored market for scientists, especially those in the mathematical field wishing to broaden their scope and interact with to other areas. At the same time, if the social sciences wanted credibility, statistics gave them an opportunity through a numerical analysis of qualitative properties hitherto analysed only in a "humanistic" sense. Gradually, the room for pure intuition, or conclusions from general observation, was receding. As said by Blackett in [1948], for "gusts of emotions".

Brief Bibliography of Fisher.

A study of Fisher's bibliographic has some interest; among the almost three hundred papers he published, he sometimes writes about two or three topics, at different levels, in the same paper. He would touch upon very technical matters in a sociology oriented paper. Nonetheless, his three main works, quite specific in their topics, are *Statistical Methods for research Workers, The genetical theory of Natural Selection*, of 1930 and *The design of experiments* published every five years from 1925 to 1935.

From 1912 the year he published his first paper, "On an absolute criterion for fitting frequency curves", until 1925, the year of his first major book, Fisher published in several journals; for example, in *Messenger of Mathematics*, where he published an article on the application of vector analysis to geometry, a topic not totally foreign to the spirit of his statistical work. He also published in *Eugenics Review, Biometrika*, and in 1918, his main paper, on The correlation between on the supposition of Mendelian Inheritance, was published by the *Transactions of the Royal Society of Edinburgh*. Some of the other publications in which his name appeared are the *Monthly Notices of the Royal Astronomical Society, Metron, Journal of Agricultural Science, Philosophical Transactions of the Royal Society of London, Journal of the Royal Statistical Society, Annals of Applied Biology, American Naturalist* and *Economica*.

He had the ability to write to his mathematician and statistician colleagues in technical journals, as much as to write to biologists and experimental workers. To make this possible, Fisher generated a unique epistemology for his writing activity, capable of linking the perception of the observations to the process of generation of knowledge. It is in this writing activity where we can observe his management of symbols and signs, his use of mathematical language and the bridges he built between the colloquial and normal language.

An interesting epistemic problem derived from the organization of his texts is shown in the following table, where CP means his Collected Papers. Numbers refers to the classification proposed by J. H. Bennett; the number in bold refers to the first time Fisher mentions the subject:

Subject	**Book/year**	***Statistical Methods* (1925)**	***Natural Selection* (1930)**	***Design of Experiments* (1935)**
Test of Significance				**1915 CP4**, 1922 CP20, 1925 CP43, 1935 CP127
Correlation Coefficient		**1918 CP11**, 1921 CP14, 1924 CP35, 1928 CP61		
Analysis of Variance		**1918 CP9**, 1923 CP32		**1918 CP9**, 1923 CP32
Design of Experiments				**1922 CP18**, 1923 CP32, 1926 CP48, 1927 CP57, 1929 CP78, 1935 CP128, 1938 CP158

Genetics and Heredity		**1914 CP3,** 1918 CP9 CP11, 1921 CP17, 1922 CP24 25 26 27 28 29, 1924 CP41, 1932 CP98 99 1935 CP131 1936 CP147	
Crop variation	**1921 CP13** CP15, 1923 CP32, 1924 CP37, 1926 CP48, 1927 CP57, 1929 CP78		**1921 CP13** CP15, 1923 CP32, 1924 CP37, 1926 CP48, 1927 CP57, 1929 CP78

From the above table we can see that the test of significance, although treated widely in his [1935], had been studied since 1915, that is 20 years before; then the concept appeared in 1925 and finally in 1935.

The correlation coefficient studied in his [1925], had been determined 7 years before as well as the analysis of variance, meanwhile the concepts related to the design of experiments had been 'in the air' since 1918.

Genetics and heredity was a topic always present in the work of Fisher. The first paper about it was in 1914, his book about genetics theory was published in 1930 and he worked on this area for practically all his life.

Meanwhile crop variations was a topic only present in a specific period of time, the first paper in 1921 and the last one in 1929, corresponding to the years when he worked at Rothamsted, nonetheless all the examples mentioned in his [1935] are related to crops and experience in these laboratories.

Another important issue to note is that analysis of variance is repeated in his books of [1925] and [1935] as well as in other papers, while the test of significance only appears in the latest and the concept of correlation coefficient only in the first. The treatment of design of experiments is not mentioned in his previous books although there are articles and papers where the topic is treated however not formalised as in his [1935].

Genetic examples were not treated in his theoretical books of statistics ([1925], [1935]), although his work on genetics is chronologically in the middle of the other works [1930]; it seems that genetics was for him an area of knowledge in which some of the main problems should be treated with other tools and arguments as well.

One can observe that the variety of types of texts and their relevance in time are meaningful to understand his work, nevertheless in this case it is also relevant to establish the position of the subject or subjects treated (in the papers) in relation with his major publications (the books) to take a perspective of when and how the consolidation of concepts are related with the development of field experience Fisher was in contact at that time. We can observe that no books or other kind of texts were written exclusively before or during some specific period, for example his stay in Rothemstand (as in the case of Blackett where no books were written during the war years), however in this case Fisher has all the kind of documents at all times.

The decision of taking into account the dates of the publishing of his books is because, as seen in the above analysis, many concepts treated in separate papers were formalised and concluded in his books. I wish to remark that in Fisher's case the epistemological platform (as the creation of concepts and drawing planes) is better seen if we analyse book/year-subject as seen in the table above.

Design of Experiments, an epistemology in itself.

The first approach in the present work is <design of experiments> and although this does not follow a chronological factor it tracks an epistemic one. Fisher's statistical work is based on his experimental method and all his contributions revolve around experimental factors and measured observations.

His widely known major breakthroughs in applied mathematics to non-scientific fields (analysis of variance and correlation coefficient) refer directly to observations, data tables and experimental evidence. Other major breakthroughs in his statistical work were also within the frame of experimental work, for example the test of significance: This concept refers to the problem of the minimum size of a sample to infer valid and significant conclusions out of it.

The reason why the test of significance is not considered as an independent breakthrough in the structure of the present work is because it is more valuable (in the epistemological argument) as part of design of experiments and also due to its influence in the other mathematical achievements. In other words, if the number of observations (size of the sample) is not representative of the total population, no analysis of variance or correlation coefficient or any other statistic is going to draw valid conclusions on the population. As the Fisher puts it:

> If the design of an experiment is faulty, any method of interpretation which makes it out to be decisive must be faulty too. [Fisher, 1935, 3]

Obviously, to question the context (in the sense of framework) in which the experiments are done implies questioning directly the function of the authority that draws the conclusions of it. In other words, the authority that administrates a "scientific reality" is creating an auto sufficient and autonomous scientific language.

That is why the analysis on the purpose of writing the book about how to design experiments and positioning it as a book to follow by experimental workers is of the utmost relevance to our case.

The purpose: internal business

Fisher's reasons to write on the design of experiments seem to be more political and strategic, in relation to the position of scientists in the organisation in which they work, than any other more academic or scientific initiative of the time. For example, the design of experiments did not come from a necessity to solve a specific problem or to develop a certain model. It seems that the experimental workers (as part of Fisher's team) had to deal with colleagues and bosses in the same organization and validate his work in order to confer significance.

> Critics who still refuse to accept the conclusion (scientific conclusion proved on experimental evidence) are accustomed to take one of two lines of attack. They may claim that the interpretation of the experiment is faulty...The other type of criticism to which experimental results are exposed is that the experiment itself was ill designed, or, of course, badly executed. This type of criticism is usually made by what I might call a heavyweight authority[5]. [Fisher, 1935, 1-2]

In a sense this is a way to uphold the work and the dignity of the research and, in consequence, of experimental workers. The italics of the word "authority" might refer, as well, to the authoritative attitudes without reason and above all from people that did not know about the experimental field and had some organizational authority in it.

> The authoritative assertion "His controls are totally inadequate" must have temporarily discredited many a promising line of work; and such authoritarian method of judgement must surely continue, human nature being what it is, so long as theoretical notions of the principles of experimental design are lacking [Fisher, 1935, 2]

The book Design of experiments was published in 1935; but to put it in perspective from a social point of view, one has to be aware that in 1933 Fisher's contributions in field of genetics led him to be offered a post at University College London, as indicated before.

Egon S. Pearson joined his father at the Galton laboratory in 1923 and followed his father's ideas in the Bayesian concepts. It is important to note that relations between the latter and Fisher were not exempted from academic incidents. In 1926 Pearson published (in Biometrika) the results of an experiment design to test Bayes' theorem. These results comprised some 12,000 fourfold tables observed under approximately random sampling conditions. Then, in his article in

5 Italics are Fisher's.

Eugenics Review of 1926, Fisher calculated the actual average of χ^2, which he had proved earlier should theoretically be 1 and which Pearson still maintained should be 3. In every case the average was close to one, and in no case near to three:

> The mean to be expected is thus 1.02941 with a standard error of only .01276. The mean value from E. S. Pearson's sample is thus lower than the expectation by a very small, but statistically significant amount; showing that the conditions of independent random sampling, though very nearly, were not exactly realised. It is hope that this example will remove all doubts as to the correct treatment of the fourfold table, and of other applications of the c^2 test. [Fisher, 1972, 96]

Thus, concluding the lack of randomness of the conditions of the experiment. There was no reply to the article. It seems that a conflict began on top of the fact that Fisher never wanted to work under Pearson's leadership (it has been said before that in 1919 he was offered the post of chief statistician under Karl Pearson at the Galton Laboratory but he chose the post of statistician at Rothamsted Experimental Station under Sir John Russell).

The two departments shared the same facilities making the atmosphere difficult due to the irreconcilable conflict between Fisher, the Pearsons and, probably, his followers. Having two authorities at the same level was not the most comfortable environment and perhaps the internal politics of the department were complex.

Although in 1930 Fisher taught courses in statistical methods at Imperial College (in South Kensington) and also at the Chelsea Polytechnic, he did not lecture on pure statistical theory, but only on the Philosophy of Experimentation, this was due to the problems with Egon Pearson. In fact Fisher wrote a letter to him stating the possibility of a problem of opinions:

> ...During the last few years a number of students have come to me under the guise of voluntary workers, with a view to learning statistical methods, especially in those points in which I have differed from your father. I think that we must anticipate that this would continue and only hope we can avoid their becoming an embarrassment. I suppose the academic ideal is that they should hear both views and use their own brains but I have not enough experience to know that this would work out well and shall be glad to know how you view this particular difficulty, which I have put before you as frankly as possible [Fisher J, 1978, 258]

Pearson answered the following lines which appear in Fisher's papers:

> I would suggest that you should not start by giving any lectures in your department in pure statistical theory. I image that the training that you have been giving in statistical methods need not take lecture form. Later on, when you are quite sure of the ground, I would like to think that you would be ready to give some lectures on your outlook on statistical theory which would fit into courses of the Statistical Department. At the same time, I have no doubt that some of your students will choose to come to my lectures. [Fisher J, 1978, 258]

In his [1935] he discredits the position of a non-competent figure in some important post, and justifies the correct design of experiments:

> …the subject matter of this book has been regarded from the point of view of an experimenter, who wishes to carry out his work competently, and having done so wishes to safeguard his results, so far as they are validly established, from ignorant criticism by different sorts of superior persons.[3-4]

The above paragraph seems to be pointing to conflicts with authorities like Pearson's son who, in Fisher's view, did not have anything to contribute to or even to say about the statistical treatments of data and conduction of experiments, as his work had been in theoretical statistics. Probably, the true reason to write a document like this was to avoid being discredited from the academic <enemies>; and above all <enemies> against his collaborators and followers; a powerful weapon in this environment was to have the "best argument" and a way to do it was the systematic presentation of results, a topic treated in his *Design of Experiments*.

In these circumstances, Fisher developed his *The Design of Experiments*, incorporating a complete section on the rejection of *Inverse Probability*, to continue with the logic of the laboratory and to finish with the uncertainty of precision in the experimental work. It is important to note that the examples are all of agricultural and biological fields, no genetics or heredity is mentioned in the entire book except for a brief note of 2 pages at the end of it [241, 242], but not taken as examples. There seems to be a vision of genetics, which puts it out of the experimental frame; I wish to insist that Fisher was interested in these topics since his years at Cambridge.

Fisher was assertive in his ideas, he was a political conservative, a traditional member of the church of England and a fervent nationalist [Fisher J, 1978, 11]. He had also read and admired Nietzsche; in fact when he was in Cambridge he formed a group to read and talk together taking in the phraseology of *Thus spoke Zarathustra*. He also read the works of Charles Darwin and became interested in genetics since then.

In fact Fisher was highly concerned with the evolution theory and sociology itself, and tried to establish relationships between qualitative properties, traditionally treated in the humanities or social disciplines, and quantitative parameters, as in the chapter VIII of his book The genetical Theory of natural selection, which has the sujestive title: Man and Society. In it he deals with human organizations, economy, history and philosophy, and end up with a quantitative analysis of human fertility related to moral and social conditions. As an example he quotes some data of the census of 1901 calculating mean and variance to infer some conditions to be observed in the human society.

> The mean number of children for this group of women, for whom the chosen conditions are especially favourable to fertility, is 8.83, but the variance is still as high as 12.43. [Fisher, 1930, 190]

The influence of the cultural environment in Cambridge at the time can be connected to a person, in our case to Fisher. The concept of "average man" compared with the "super man", widely discussed by Ortega y Gasset and Nietzsche respectively might have also been an influence; for example, Nietzsche sustained in his book *Beyond Good and Evil* that the process of education most be regarding as the continuation of the process of breeding, as the two activities had the same object: to create beings more capable of surviving in the struggle of existence, an statement that could be connected to Darwinism and fro him to eugenics.

The notion of mathematics for Nietzsche was that it falsified "reality" because it relied on presuppositions such as the existence of a precise straight line and how the process of conceptualisation was achieved a-priori.

A good argument to support this view of mathematics in the field of statistics is to reject the principle of inverse probability and base the science in the empirical knowledge, which goes in line with Nietzschean thought. The phenomenologist philosophy was also a cultural factor at that time, and Fisher, being interested and in contact with philosophical ideas, like Nietzsche's ideas, was concerned about this kind of approach; we do not think that Fisher could have read Husserl's work, however, the coincidence of concerns is present at least in the mathematical approach,

Putting together the reasons that motivated Fisher to write are exposed in the same text as seen above, but also his historical situation and cultural influences might have played an important role to develop such ideas.

Elements: the generation of information out of simple data.

The elements of the design of experiments are explained throughout the entire book in many levels and steps and sometimes even cryptic as not defining ideas and introducing not conceptualised elements, for example he mentions the conclusions and the validity of data since the beginning, and the observations and recording of data at the end, nevertheless he follows a <general epistemology> which is not the same as in other successful experimental scientist like Blackett. This is illustrated in the diagram on the next page.

It is essential to be aware of this comparison because differences in concepts and terms are not purely semantic, but epistemologically, ontologically, semiotic, and hermeneutically different, for example the concept of experiment is defined, applied and interpreted in different ways, also they do not occupy the same step in the epistemic processes. These factors meant the difference between successful applications, implementation and transcendence not only in scientific fields, such as biology or medicine but also in the power spheres like military and ministerial policies. Therefore factors like these affected the generation of significant knowledge in the area of applied mathematics.

Seen in this way, the scientific method does not need of ontological reality, but operates and produces its results in the experiential domain of the observer. The representation in the classical ontological terms puts language in a hierarchy of intellectual resources

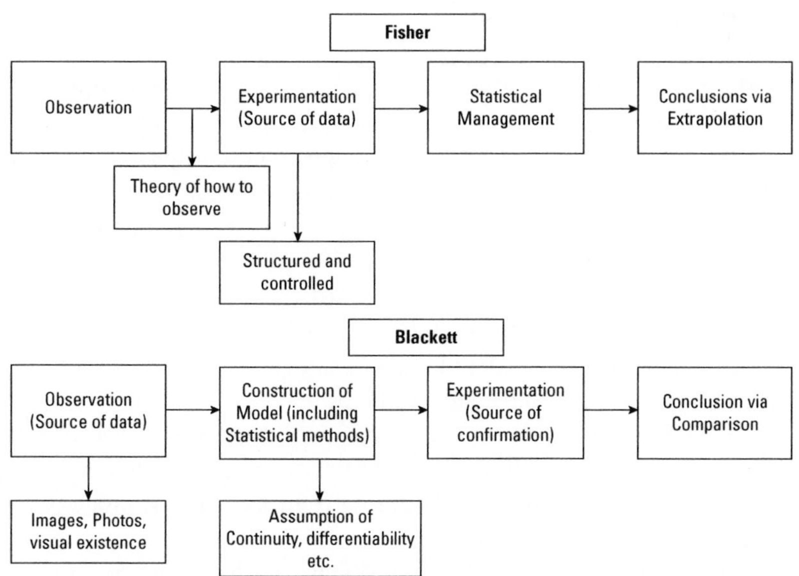

in which such an activity submits its legitimacy to an "objective" knowledge which is also validated institutionally by the reality of the world "reality" based in the Kantian "Ding an Sich".

Our epistemic proposal since the beginning is consistent with an ontological level reflected in all the cases analysed for this work (Blackett, Fisher and Kantorovich). In order to analyse these cases, our epistemic proposal can be seen from the semiotic point of view as referred to a notion of discourse (the classical semiotics came from a linguistic discourse) or narrative in such as "ontological standardisation", in which the absence of analysis concerning the problems of Eco's dictionary and the Encyclopaedia (see [Eco, 1979]), might generate in each scientific writing. The notion of a writing activity in the Piagetean sense (as a necessary component of knowledge) appears, in the case of Fisher and Blackett, in the translation from technical language to colloquial one and also transforming situations and objects of the real world, like observations in laboratory into a report or experimental survey, paper or book in which a given narrative as equivalent of knowledge is placed (see [Glaserfeld, 1995, chapters 3 and 4].

First epistemic steps: Observation-Experimentation.

Observation is the basis for Blackett's and Fisher's work; but it has different views and approaches. For Blackett it is the source of information, many times shown via images, photographs or tables of data, while in Fisher it is the basis for formulating a theory (a prescription) in order to conduct an experiment which is the true source of information. For Fisher this step is the most important of all, his arguments revolve around the problem of the correct observation to infer correctly out of some sample, in fact his [1935] is dedicated to show how a set of well-ordered and controlled observations should be treated, organised and managed statistically.

His questions about nature are not in the form of experimental operations, but in a way that only statistics could constitute an answer. His questions on the observations are more in the form of propositions than in the form of questions.

This leads us to the issue of <problematisation>, in this case it is not in the question or the explanation (like in Blackett's), but in the direct inference drawn from the evidence shown that triggers the enquiry. The essence of problematisation is directly in the method of inference:

For example this can be seen precisely in the case of an experimenter who wants to compare the effects of two types of fertilizers, A and B, on certain crops. He poses himself the question: What is the difference between the yield of plot A under this fertilizer and that of the plot B under the other? But his "question" comes in a form of proposition that reflects what he expects out of it. This is what Fisher calls "null hypothesis".

The null hypothesis is a proposition about the expected result from the experiment, and its form should be precise, i.e. free of vagueness and ambiguity, because it must supply the basis of the "problem of distribution" of which the test of significance is the solution.

> This hypothesis, which may or may not be impugned by the result of an experiment, is again characteristic of all experimentation... A null hypothesis may, indeed, contain arbitrary elements, and in more complicated cases often does so. [Fisher, 1935, 18-19]

The essence of experiments is not to confirm the null hypothesis but to reject it. This hypothesis is related directly to the observation and the number of elements generated by the set of observations of that class:

> By increasing the size of the experiment, we can render it more sensitive, meaning by this that it will allow of the detection of a lower degree of sensory discrimination, or, in other words, of a quantitatively smaller departure from the null hypothesis. Since in every case the experiment is capable of disproving, but never of proving this hypothesis, we may say that the value of the experiment is increased whenever it permits the null hypothesis to be more readily disproved. [Fisher, 1935, 25]

Experiments are not to corroborate, but to create the basis for the whole study and explanation of it. This is contrary to Blackett, who used the experiment to corroborate a theory that was a-priori done or figured out.

Experiments are in Fisher the source of raw material; in his [1935] he emphasises on the first two steps: observation-experimentation as the basis for the whole statistical (scientific) study, these steps and the whole design of experiments have to be solidly based: he explains how to avoid the biases and errors in observation that are for him probably the most important thing in the generation of information.

Fisher is concerned with what we have called the epistemic problem of information production. This was achieved via an interpretation while observing, which could be any, depending on the point of view of the experimenter, the variables he imagines, or the approach to the already given information. In Blackett's case he is not concerned about that, illustrated in an example, in Operational Research the way in which the soldier gets the information does not matter, in Fisher's case he has to be witness of his own observation.

Validity.

One of the most important concepts in the design of experiments is validity. Fisher establishes a level of confidence in the result of the experiment of 5%, i.e. 1 out of 20 tries. This is explained in his article of 1926 that appeared in the Journal of the Ministry of Agriculture of Great Britain:

> If one in twenty does not seem high enough odds, we may, if we prefer it, draw the line at one in fifty (the 2 per cent point) or one in a hundred (the 1 per cent point). Personally the writer prefers to set a low standard of significance at the 5 per cent. A scientific fact should be regarded as experimentally established only if a properly designed experiment rarely fails to give this level of significance [Fisher, 1972, 85]

In the early 1920s the number of replications of each treatment was usually three or four. It was an empirical compromise. Every additional replication added some factors to the accuracy of the estimate error, but if as many as 10 or 12 different varieties were compared, restrictions of experimental space and labour had to be considered .

Test of significance.

The test of significance is exemplified with the example of the lady who can distinguish between a cup of tea which was prepared with milk added first and one which had the milk added after. The incident happened when one afternoon, at tea time, Fisher drew from the urn a cup of tea and offered it to the lady beside him, Dr. B. Muriel Bristol, an algologist. She declined it, stating that she preferred a cup into which the milk had been poured first. "Nonsense, surely it makes no difference", said Fisher, but she maintained that it did. It was William Roach[6], a colleague working on insecticides who suggested testing

6 He married later Ms. Bristol

her[7]. They prepared the tests for the experiment which formed the second chapter in Fisher's [1935], entitled "The principles of experimentation illustrated by a psycho-physical experiment", beginning as follows:

> A lady declares that by tasting a cup of tea made with milk she can discriminate whether the milk or the tea infusion was first added to the cup. [Fisher, 1935, 13]

The first question was concerned with how many cups of tea should be used for the experiment, then in what order should they be presented, and most important what kind of conclusions should be drawn from the results, let's say a perfect score of all the cups identified correctly.

The experiment consisted in this: "Mixing eight cups of tea, four in one way and four in the other, and presenting them to the subject for judgement in a random order" [Fisher, 1935, 13].

In order to get the 20% of confidence one has to calculate using combinatory analysis the number of cups of tea that should be used for the experiment. To achieve 20% it is not enough to have 6 cups (3 of one kind and 3 of the other) as the perfect score is exactly 20% of the total possible cases, thus any result can be attributed to chance one out of $20 = C_3^6$. In the case of 8 cups (4 and 4), achieving a perfect result, i.e. 4 correct identifications, is 1 out of $70 = C_4^8$, more or less 1%, however 3 rights and 1 wrong are above the 20% ($\frac{16}{70} = 22\,\%$).

Of course there could be some other variables not controlled by the experimenter or simply not contemplated in the experiment, that could be affecting its result, like the temperature of the milk, or the tea, or the size of the cups, or the order of preparation (some of them can be slightly colder than others) etc. so the subject, in this case the lady, can argue that most of the time she can identify them. In this, and in order to draw a valid conclusion, one has to grow the experiment or in a way to repeat it.

7 See [Fisher J., 1978, 134]

Although we do not know for sure if the lady was right or not, the experimental method for investigating was positioned as one of the main statistical tools.

Repetition.

An experiment can "grow" or be increased, in the above paragraph, "By increasing the size of the experiment", means two things: on the one hand increasing the number of observations or repetitions but not the number of variables to be controlled, not the area in which some fertilizer is tested and not the time in which the experiment is held, for example having 1 out of five trials or 2 out of 10 as a limit for the significance of the inference, i.e. if the lady identifies 8 times out of ten the case of 3 rights and 1 wrong, then one can draw a valid statistical conclusion; and on the other it means to increase the size of the sample, in the case of the cups of tea one can choose 12 cups and the successful classification can be 5 right and one wrong drawing valid conclusions: 37 favourable cases out of 924 (within the 20%).

All these variables are already determined in the form of the null hypothesis, and the number of observations is related directly to the size of the sample to be observed. "We can render it more sensitive, meaning by this that it will allow the detection of a lower degree of sensory discrimination, or in other words, of a quantitatively smaller departure from the null hypothesis" refers to the degree of confidence that can be measured by the size of the sample, but always related (not depending upon) to the null hypothesis and not in the size of the population about which the conclusion of the experiment will infer.

> The structure of the experiment is determined when it is planned, and before the content of its results, consisting of the actual yields from the different plots, is known. [Fisher, 1935, 58]

Only to note a lack of epistemology from the test of significance, this is concluded paradoxically from a frequentist point of view, calculating combinatorially the number of trials necessary to achieve the 20% above mentioned and giving the equally probable property to assumptions like "pure chance events"

Randomisation.

To increase the sensitivity of the experiment, it is relevant to contemplate the process in which the cups are presented to the subject, for example if it is presented in some special order to prove the degree of sensitivity of flavours in the subject. The randomisation of the process can be done as "by the toss of a coin" so that each treatment has an equal chance of being chosen. This process was given first in the organization of the samples involved in the experiment.

Randomization gives the objectivity value to the experiment trying to avoid that the events would have been done by chance or biased by some bad design. This was more useful in more complex experiments involving agricultural factors using the <Latin squares>; showed by Fisher as follows:

> When the idea of effecting an elimination of errors due to soil heterogeneity in two directions at right angles was first appreciated, the necessity for randomisation in experimental trials was not realised. In consequence, certain systematic arrangements were adopted. On of these, which may be called a diagonal square, is shown below:

A	B	C	D	E
E	A	B	C	D
D	E	A	B	C
C	D	E	A	B
B	C	D	E	A

> [Fisher, 1935 85-86]

In the above table one can see that the diagonals are motif of bias, as in the following table of a systematic arrangement of 16 plots in 4 blocks to test 4 different types of substances, the columns have the same treatment.

A	B	C	D
A	B	C	D
A	B	C	D
A	B	C	D

The bias due to the type of soil, sun water, inclination, and other many variables can affect the experiment and lead to wrong conclusions. This was fixed via combinatorial methods ending with the following distribution:

> Consequently, an improved systematic arrangement due to Knut Vik has been widely used. In this each row is moved forward two places instead of one, so that the arrangement is as follows:
>
> | | | | | |
> |---|---|---|---|---|
> | A | B | C | D | E |
> | D | E | A | B | C |
> | B | C | D | E | A |
> | E | A | B | C | D |
> | C | D | E | A | B |
>
> [Fisher, 1935, 87]

Finally in the case of the cups of tea, the experiments can be modified in such a way to achieve more sensitive results:

> Thus we might arrange that 5 cups should be of one kind and 3 of the other, choosing them properly by chance, and informing the subject how many of each to expect. But since the number of ways of choosing 3 things out of 8 is only 56, there is now, on the null hypothesis, a probability of a completely correct classification of 1 in 56. [Fisher, 1935, 27]

Although it seems that theoretical analysis deals with the design of experiments, one can situate and determine it within this philosophy of experimentation as one of its steps, i.e. design of experiments is an epistemology in itself and includes analysis as one of its parts.

The step of observation is crucial because it relies in this part the discussion on the inverse probability versus the experimental frequency-evidence and it is always related with the objective knowledge and the concept of truth.

In 1919, when Fisher began his work in Rothamsted, his first task was to see what could be learnt from statistical analysis of the records of experimental and observational data collected over a number of years[8]. In the years of 1914 to 1919 agricultural scientists were called to Rothamsted to deal with problems of food production during the

8 See [Fisher J, 1978, 141]

war. Sir John Russell was director of the laboratory since 1912 making expansion in the laboratory: in 1913 the physics department began with Sir Bernard Keen, the department of Bacteriology opened under Sir Gerard Thornton, the entomology and organic chemistry departments began to study insecticides and problems of infestation derived from the needs of the war and the classical departments of chemistry and botany were strengthened. The department of protozoology was recreated and the department of plant pathology was concentrated here after being divided one part in Kew and the other in Manchester under W. B. Brierley (mycologist) and A. D. Imms (entomologist). Imperial College set a laboratory of plant physiology in Rothamsted and in 1923 the Bee Section of the Ministry of Agriculture was transferred there[9].

In this environment a 29 year-old scientist came to the active work in the laboratory. To compare, in the case of Blackett he was 40 years old and a consecrated figure in the academic and scientific fields when he began to work actively in the service during the war. Fisher did not have the authority of a figure such as Blackett to work with PhD students recommended from the colleges, but he did work with experts in their own fields and high and respected figures in the academic world.

So where is the scientific basis in the process?[10] In Blackett's case it is in the proposal of the model itself. This model was conceived in the process of observations, or as called by Blackett "taken out from experience", this model is <coded> in mathematical language and algebraically manipulated; when the logical machinery of mathematics is finished there is an expected result, which is confirmed in the experimentation.

Fisher performs radically differently: he realises that before observations and experimentation, there are some theoretical considerations done by the observers (null hypothesis), the interesting question here is whether there was something to expect from the experiment, as in the case of Blackett, or whether there was something

9 See [Fisher J, 1978, 96]

10 Meaning science not the activity of dealing with reality, but the theoretical (narrative) treatment of facts and concepts

to expect from the primary observation. These are two views of constructing knowledge from experience.

Throughout his work, this is a problem that Fisher attacks, for instance emphatically denying the principle of inverse probability (which is based on an a-priori assumption totally independent from experience, observation or previous information). But what place has the expectation, which was independent of any observations or theoretical basis? In fact the relation between "the expected" and the "truth" was crucial; this is reflected in the passage of "Concerning the unsuccessfulness of Experiments" by Robert Boyle, written in 1673, which Fisher quotes in the beginning of his book of 1935 [Fisher, 1935].

> …you will meet with several observations and experiments which, though communicated for true by candid authors and undistrusted eye-witnesses, or perhaps recommended by your own experience may, upon further trial, disappoint your expectation, either not at all succeeding constantly or at least varying much from what you expected [Fisher, 1935, xi].

This passage refers more to Fisher's concern about the importance of experiments but also reflects his point of view on "expectation"; What is the meaning of "from what you expected" in the above quotation? The scientific conclusion or <guess> has been done before the experimental evidence, it is based on observations, but the true scientific process begins with the experiment, as one of its objectives is to base (not to confirm, like in the case of Blackett) such hypothesis made by <theoretical science>.

Science for Fisher is not an a-priori process separated from the "well-ordered" and controlled series of observations called experiments, but it is an a-posteriori conclusion based on it, nevertheless out of the observation there is some kind of knowledge an "expectation", i.e. there is knowledge far before the experiment is conducted and probably before observations are taken, this knowledge is referred by some as Scientific Knowledge, but for Fisher the true science is more concerned with the internal epistemology of the reasoning about the outcome data from experiment, than with dealing with the <physity>, (<physity> understood as the physical stable system subjected to perception, in other words, its physical nature, (see

[Canaparo, 2000]). or any system or machine or process of <reality>; this is a clear distinction from science and technique: science is the qualitative treatment of dealing with the quantitative analysis.

Observation is then, the first approach to attack problems of real life, and a set of controlled and ordered observations is called an experiment. Experiments are derived directly from observations and the expectation is translated from a qualitative property to a quantitative variable, (most of the times a statistic: mean, variance, etc.), for example, from "The people in the army are taller than in the navy" to "the expectation is that the mean is greater than 5.11 ft.".

Observations lead to the idea of populations and experiments to samples from populations. Fisher is aware of the ontological level in which he works and says:

> If we have the records of the stature of 10,000 recruits, it is rather the population of statures than the population of recruits that is open to study. Nevertheless, in a real sense statistics is the study of populations, or aggregates of individuals, rather than of individuals. [Fisher, 1925, 2]

As we will see in this work, Fisher is more concerned with the idea of "experimental science", without which it is not possible to understand his point of view of science, progress, mathematics and its applications to other sciences. This is clearly visible in the analysis of the methods (statistics) and experiments performed in his field (basically biology, agriculture and genetics), nonetheless he mentions further applications of his methods in other areas, like physics or economics. For example when speaking about the study of statistics he says:

> Scientific theories which involve the properties of large aggregates of individuals, and not necessarily the properties of the individuals themselves, such as the Kinetic Theory of Gases, the Theory of Natural Selection or the chemical Theory of Mass Action are essentially statistical arguments [Fisher, 1925, 2]

And

> Since the method of measuring information, which has been illustrated [design of experiments], is applicable to data of all kinds, it is only necessary, in order to

> ascertain how much information is utilised by any proposed method to determine the sampling distribution of the estimates obtained by the method from the quantities of data of the same value as those observed [Fisher 1935, 244-245]

Observation evolves in the epistemic process and becomes from the source of expectations, guesses, or the basis for experiments, to an <event>. An <event> is an element of the experiment, and of course it varies according to the kind of experiment and its conditions, for example a) an event is a result (head or tail) from a toss of a coin in the context of recording the parameters related to the tossing itself, b) an event is a chicken born female in the context of measuring the sexing linkage in a genetic experiment, or c) an event is a difference in the size of a seed in the context of agricultural cross-fertilized experiments. The first two examples might be binary while the third could be numerical (and the statistical treatment might be different, in one case deducing a distribution of events (counting the number of occurrences) and in the other calculating statistics like mean or variance)

In Fisher's point of view, science begins in the experience and recording of observed parameters of reality. As an example of recording and tracking information, the following working papers illustrate how the calculations of averages and the elaboration of diagrams are active part of the analysis. [Fisher Papers, Cambridge University, not catalogued yet].

and perpendicular

(x')

①	②	③	④
−3	−1.5	−6	+6
+5	−6	+2	−4
−6	−4	0	−½
−2.5	−2	+2	−1.1
0	−4	−6	+3
−1.1	−2	+7	−2.5
−1	0	0	+5
+2	0	+5	+3
−4	+3	+3	+4
+.5	−15	+10	−2.5
+5	+3	−16	+3
−10	+3	+3	+8
	+2	0	−0.5
		+3	+1
			−3
(11.5)	(13)	(13½)	(14)
+ 10.5	+ 11	+35	+ 33
− 27.8	− 35.5	−28	− 14.3
Sum 38.3	46.5	63	47.3
diff −17.3	− 24.5	+ .7	+ 19.7

Total 16.9

① and ③ } $\overline{x'}$ $\frac{-17.3 + .7}{25} = -0.41$

② and ④ } $\overline{x'}$ $\frac{-24.5 + 19.7}{27} = -0.18$

} P_x 0.434

Conclusion $\overline{x'}_{①③}$ and $\overline{x'}_{②④}$ are both zero within probable error.

i.e. no direct effect of mag: field.

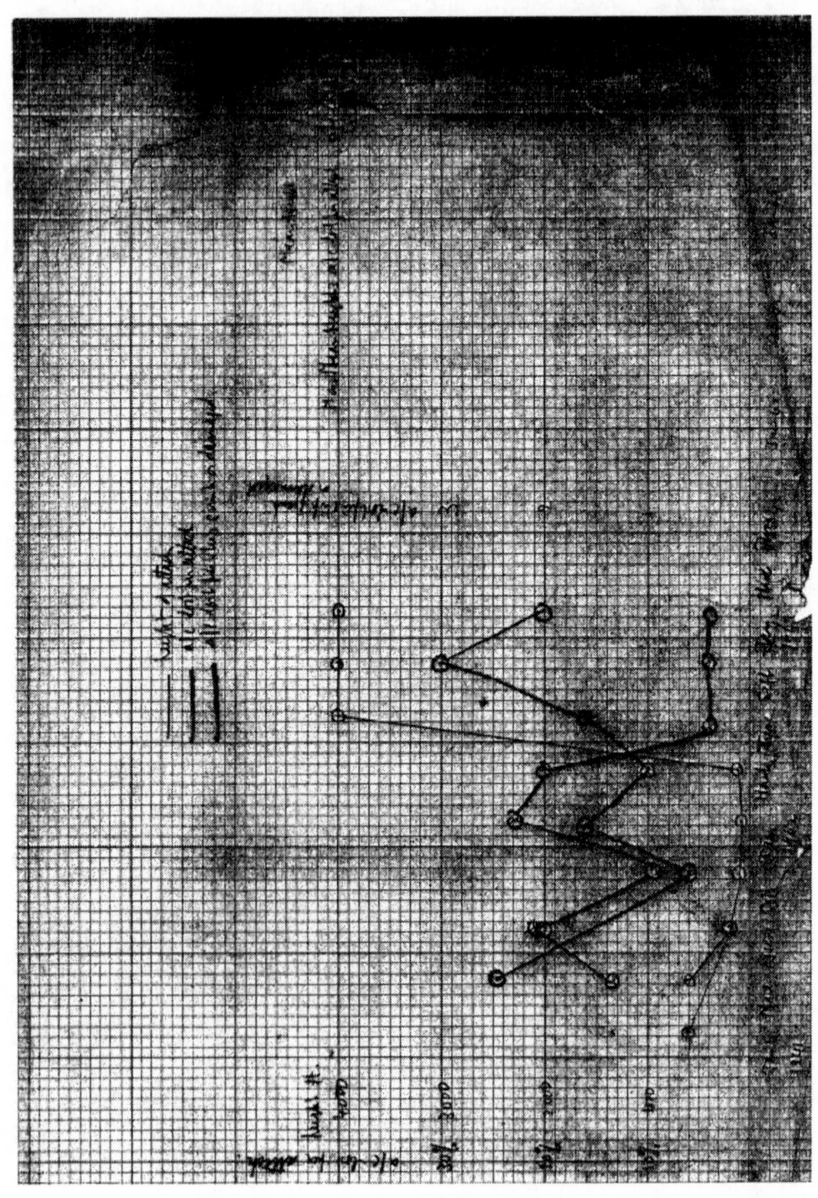

But his idea of science is developed further, and transcends the purely scientific sphere; Fisher also has an idea of science as a whole, a science linked with other fields of human activities. The relation between nation, subject, and science, in fact, is also present in Fisher.

Science has its own territory, science lives not in a country or in a person, but has its limits and its language, has its land and its rules, this is reflected in an attitude; in Fisher's case in the attitude of objective observations, but how the observations can be objective if they are done by a particular observer? The answer is complex but the first approach to objectivity is via numerical language, which serves as signs to communicate the reality of the world.

Design of experiments created a territory, this territory newly founded by Fisher, is a territory of activities more than of theories, nevertheless the conclusions are the attempt to communicate with other territories; it is precisely in this process that "statistical science" is of the utmost importance as statistics is applied in biology, genetics, physics, and even in psychic fields [Fisher, 1972, 388],

This is a new point of view specially referring to other sciences: A new way of reading a <expecting horizon> (horizon d'attente). The notion of 'Expecting Horizon' is taken from the prominent work of P. Ricoeur of 1984, Temps et Recit, and it is the space generated by the expectations of a certain present projected in the immediate becoming.

Summarizing so far, observations or experiences instead of experiments, are basis of our opinion, but experiment (the set of experiences or observations) are the basis of information for the scientific analysis which relies on the statistical treatment of the data; that is why one has to prepare the set in the most accurately way possible.

Hermeneutics.

An event or observation has a duality in its meaning, this hermeneutical process is described as follows: on the one hand it is the first approach to a real fact and on the other is an element of the experiment called <event>, the actual interpretation is also dual, a) the scientist has an <active interpretation> (as an interpreter of music that has a music-score in front of him/her) that takes action in the fact of interpretation itself, decides the emphasis and moreover, evaluates to transform the original source into another: the next step in the process and b) a <passive interpretation> (as an interpreter of the Bible), that is critical and is discursive, this kind of interpreter does not perform as the other one, but maintains at margin in the describing of events to uphold some objectivity always necessary in the discourse of experimental science. It sounds paradoxical or possibly contradictory that these two opposite views are able to coexist in the context of scientific work. How is this possible? The following diagram illustrates it:

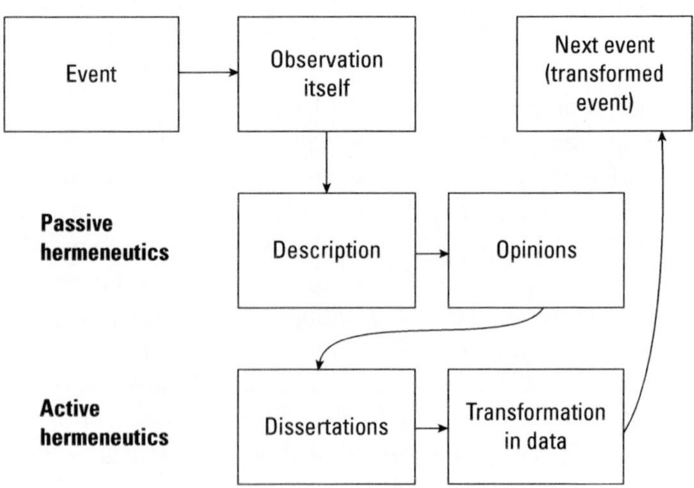

This is given first in the mere act of observing which is an hermeneutical fact by itself, then in the description that leads to opinions about the observations, these are connected with a dissertation about i) future outcomes or ii) causes of the event, then observation is translated into the formalization called recording of data in tables, so far, these observations are transformed into a more abstract level (the quantitative member of a class) in order to prepare the next hermeneutical step whose object is the next event (transformed event): the translation to a mathematical language is in fact the changing from passive to active hermeneutics; it is the step that allows to move in the process of transforming ideas into meaningful knowledge; in the diagram the steps of active hermeneutics belong more to the experimental than to the observational part. The comparison of this double denotation is in [Fisher, 1935]:

> Perhaps it should not, in this case, be called an experiment [talking about the validity of estimates] at all, but be added merely to the body of experience on which for lack of anything better, we may have to base our opinions. [Fisher, 1935, 39]

And

> Men have always been capable of some mental processes of the kind we call "learning by experience". Doubtless this experience was often a very imperfect basis, and the reasoning processes used in interpreting it were very insecure; but there must have been in these processes a sort of embryology of knowledge, by which new knowledge was gradually produced. Experimental observations are only experience carefully planned in advance, and designed to form a secure basis of new knowledge; that is, they are systematically related to the body of knowledge already acquired, and the results are deliberately observed, and put on record accurately. [Fisher, 1935, 9]

Hermeneutical analysis examines the energies of interpretation and the disciplines of understanding in the process of creating new and meaningful knowledge. Scientists like Fisher or Blackett seen as interpreters become translators of languages, and decipherers and communicators of meanings. But in order to do so they make assumptions and work on objects constructed from their experience, which for them are real, as real as the existence of a tree in a garden or the existence of the cubic root of 8 can be. Considering the reality of objects leads us to the ontological discussion of observations.

Ontology of observations.

For Blackett observation is visual, the existence of it relies on the image itself, and as in operational research the photographs acquire a level of reality in the physity of the world after the process is completed, i.e. what it is seen in the photos is considered real.

In Fisher's case, reality of observations is the logical reality, that is, it is real if it is an element of a set with certain restrictions: for instance if it is not empty within a class of elements, then it exists as an observation. Observations, in order to transcend to the next epistemic level must have a level of existence, and this is given by the membership to a class or set; taking the examples above to illustrate this point, we have that the event of a toss is an element of the set of tosses, in fact to calculate a probability in the usual way one must have a set of events (the possible cases) and another class of events (the favourable cases), but they must exist as part of a class otherwise the context of all epistemological formality is lost. i.e. we do not consider the event- in the context of a toss- of a coin falling on the edge, or not falling because a bird took it while it was in the air, but we only consider the events head and tail.

Semiotics.

Note: "In this section words that appear between triangular brackets are meant to have a special definition, for example <argument> would mean a sign which, for its interpretant, is a sign of law. An argument is a sign, which is understood to represent its object in its character as sign. [Peirce, 1986a], which is the definition by Pierce of a class of sign. Otherwise argument (without brackets) will have a normal connotation, like a point or series of reasons presented to support or oppose a proposition, or a summary of the plot or subject of a book or a process of deductive or inductive reasoning that purports to show its conclusion to be true or formally, or a sequence of statements one of which is the conclusion and the remainder the premises, another case is when specifically indicated, as in the case of mathematics. Where it would mean an element to which an operation or function, applies to its independent variable".

As said before, the set of observations is the essence of experiments. Observations refer to populations and sets of observations (already in a class) are called samples and refer to subsets of the population. This process of references and creation of concepts and entities are rich in the generation of signs, which give power to the arguments and concepts in the scientific process, above all in the writing stage. Semiotic analysis will give the basis to understand the power of the writing activity and the generation of ideas as symbols and icons even in our modern culture, affecting our methods of production and management and above all for the generation of knowledge.

These signs sometimes are connotative (coded), other times are denotative (non-coded), the images that scientists have lie in the signs of the objects that they study. Fisher takes advantage of signs generated by iconic massages well connoted in society and launches arguments to argue in the very questionable field of probability and statistics.

The semiotics of observation is first shown in the relation between the interpretant (in this case the opinions) and the signs associated with events in this stage (the observation itself), then by the relation with the object that they refer to (events). In one case observations are what Peirce calls <dicents>[11]; they are far from a theoretical or abstract proposition, they are descriptive in the sense above mentioned, they are not indexed as a, b, or c, but they are referring to images of reality, for example the observations as signs worked on the psycho-physical experiment:

> A lady declares that by tasting a cup of tea made with milk she can discriminate whether the milk or the tea infusion was first added to the cup. [Fisher, 1935, 13]

11 A Dicent Sign is a sign, which, for its Interpretant, is a Sign of actual existence. [Peirce, 1986]

And continues with the same image until the end of the experiment. Its reference is always to the description of the object taken from "reality", in this first epistemic step (step of observation) descriptive images are not changed.

To compare, in the case of Blackett observations are <icons>[12] , or signs that have direct resemblance to the object with which they have reference: the photograph itself. The object of study is the sign and the sign generates knowledge by itself. In Fisher's case icons do not exist, but the observation is immediately transformed by the interpretant into a <rhema> (on this Peirce's concept, see later in this paragraph). In this step (Observation) probably the most relevant aspect is this, because it gives an open possibility to the observation, and is linked with the interpretant (sign in the mind) more than with the object itself (like in Blackett's case) or sign itself. This generates some other signs, like events, experience, or class; the concept of <rhema> is explained by [Eco] when discussing Peirce's work:

By rhema or predicate we understand a propositional empty form, which could have been drawn cancelling certain parts of a proposition, letting an empty space in place [Eco, 1995, 36[13]]

Observation and also experiment are <rhemas>, because they are in themselves concepts that cover arrangements, management, control and "scientific" assumptions. Moreover, the stage of experiment is probably the most important of all under this view, as it generates tables of data, quantitative propositions, statistics, samples and probabilities which as signs can be interpreted as <legisigns>[14], <symbols>[15] <indices>[16] and <arguments>[17].

12 An icon is a sign fit to be used as such because it possesses the quality signified [Peirce, 1986]

13 "Per rhema o predicato intendiamo una forma proposizionale vuota quale avrebbe potuto essere derivate cancelando certe parti di una proposizione, lasciando uno spazio bianco al loro posto, la parte eliminate essendo tale che se ogni spazio vuoto fosse riempito con un nome proprio una porposizione (anche se priva di senso) ne serebbe ricomposta" [Eco, 1995, 36]

14 A Legisign is a law that is a Sign, [Peirce, 1986]

15 A symbol is defined as a sign which becomes such by virtue of the fact that it is interpreted as such [Peirce, 1986]

If probably the most important characteristic of observation is the <rhema> , at the same time, observation is an <index>, observation works as a "symptom of an illness" in medical semiology. What one can see in a symptom is a sign of other event (the illness) and the scientist tries to figure out the objects associated with those signs (causes or explanations).

Before analysing the semiotics of the experiments, it is convenient to show the following diagram that connects signs objects and interpretants in the process of the two stages (observation and experiment):

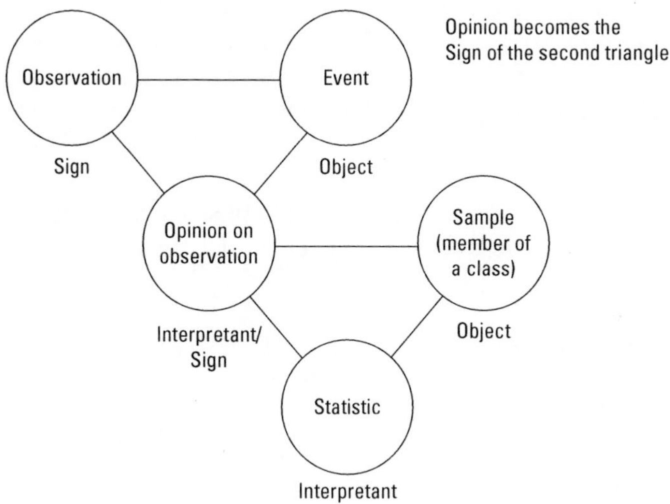

Observations, already as <rhemas>/<indices>, become an object of the second stage which is the experiment. In this case I will refer to the above (hermeneutical) diagram, to analyse all the components of this step.

16 An index is a sign fit to be used as such because it is in real reaction with the object denoted [Peirce, 1986]

17 Explained above.

The sign is related to the primary object: the observation refers to an object and the observation itself is the sign of it. As stated above, observations can be two or more types of signs at the same time. The notions and opinions generated by the observation are the interpretant as in the action of the observer, but this does not generate any other significant sign until the process goes to the "active hermenutics", where dissertation and transformation of data are generated. The dissertation is the sign, associated with an object, in this case a sample (an already set of events, and the interpretant is the statistics, which are now becoming more abstract, as they are numerical propositions. These signs can be seen as <arguments> in the form of mathematical propositions and <legisigns> as laws in certain <ground>. The concept of <ground> is crucial for the discussion, as it gives the basis for the analysis of meaning; in this case is explained by [Eco, 1995] as follows " the 'ground' of a sign is its own connotation and its own attributed character"[18].

Semiotic analysis gets a little more intricate when we try to establish in what sense the <ground> (and the meaning) differ form the interpretant. This leads to the concept of <dynamic object> and <immediate object>:

> The ambiguity vanishes in any case if it is considered that the notion of "ground" serves to distinguish the dynamical object from the immediate object, while the interpretant serves to stabilise the relationship between representamen and immediate object.[19]

And finishes:

18 "Il ground di un segno è la sua connotazione e il proprio carattere attribuito" [34]

19 La ambiguità scompare in ogni caso se si considera che la nozione di ground serve a distinguire l'Oggetto Dinamico (l'oggetto in sè in quanto abbliga il segno a determinarsi alla sua rapresentazione) dall;Oggetto Immediato, mentre l'interpretante serve a stabilire la relazione tra representamen e Oggetto Immediato [Eco, 1995, 31]

Naturally to describe the immediate object of assign, one has to appeal to the interpretant of that sign[20]:

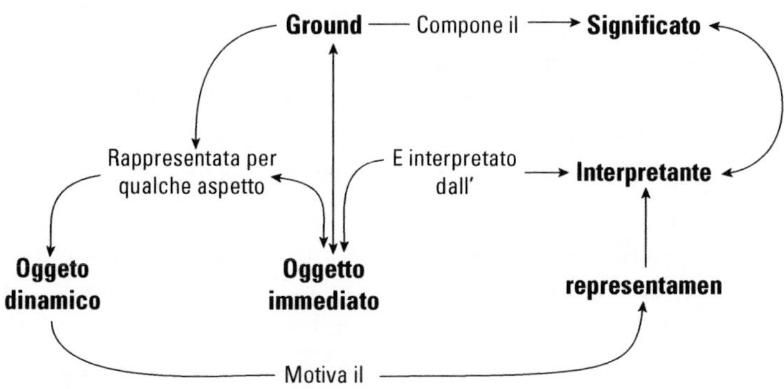

[Eco, 1995, 32]

In the filed of experimental science connected with applied mathematics (as in the cases of Fisher or Blackett), the construction of knowledge and reality is achieved by the application (even intuitively, in the authors mentioned above) of semiotic and rhetoric methods to scientific texts. That is the reason why the linguistic and language analysis can throw some light, especially on this matter.

Language, Literature and Linguistics.

Observations in agricultural and biological matters had been made since ancient times; however, in his work, Fisher's have an attempt to supply theoretical arguments to explain, and also to give conceptions, on the mechanisms that affect these agricultural processes.

20 Naturalmente per descrrivire l'Oggetto Immediato di un segno non si può che ricorrere all'interpretante di quel segno

In other cases, the observations had been there for years, as the information on fertility by Galton quoted by Fisher in his [1930, 230], or the Family records of 19th century analysed by Fisher [Cambridge University Archives, not catalogued yet].

If there had been observations and recorded information of events, what was the step forward? The breakthrough was a novel interpretation of the results that brought a decisive progress[21] to the field through a process of analysis that changed its internal methods and epistemology.

The approach in this section is to identify the structure of the texts; as a way to explain them and identify the determination of the work.

As [Barthes, 1985] says, the approach is a matter of analysis in the game of significance within writing.

> The modern writing is a truthful independent organism which grows around the literary act, it decorates it with a value strange to its intention, it involves it continuously in a double mode of existence and superpose a set of words, and opaque signs which carry in it a story, a deal... often divergent, always inhibiting the form[22].

Most of Fisher's papers are written in the present tense, and most of them in the first person singular. Although he deals with "historical" records, he actually describes the situation as a report of an experiment, and it is in this process of acquiring perceptions and relating representations between reality and knowledge that language becomes a major tool. As said by [Gross, 1996, 205] "Science is less a matter of

21 Decisive progress means the quantitative treatment of data in areas considered humanities or of social studies.

22 L'écriture moderne est un veritable organisme independent qui croît autour de l'acte litteraire, le décore d'une valeur étrangère à son intention, l'engage continuellment dans un double mode d'existence, et superpose au contenu des mots, des signes opaques qui portent en eux une histoire, une compromission ou une rédemption secondes, de sorte qu' à la situation de la pensée, se mêle un destin supplémentaire, souvent divergent, toujours encombrant, de la forme [Barthes, 1985, 65].

truth than of making words"[23]. This contact between truth and the way to express it is where the work of Fisher is located: from observation and real experience (experimental evidence) to the theoretical conclusions that explain the reality that was initially observed.

A relevant question to pose in this section is what is this text by Fisher about? Is it about mathematics, or statistics, or genetics, or experimentation? Is it a technical book, a textbook, or a general book? Are his reports and results part of his work/duty in the field or genuinely to contribute to the growth of the organization and institution?

A theory or any particular area of knowledge as [Canaparo, 2002] says, is defined by a complex interpretative, conceptual and psychological set of constraints. These activities are a problem of knowledge directly related to a theory of writing.

Referring to observation and experimentation Fisher speaks in the first person pronoun; but who is the first person in Fisher's work? The first person is the observer. In Fisher's language the observer is an active part of experiments; he manages and constructs knowledge from his particular point of view

The determinations of our knowledge are not (as seen in the above sections) autonomous or objective, but on the contrary, they almost immediately pass from that stage of objectivity to be elements in the construction of a theory to make possible the observation. As stated in the ontology section, existence is to exist logically.

Experiment, and in general statistics, is not about the phenomena, but about the evaluation of the language surrounding it. As in Blackett's case, the use of metaphors to describe phenomena, appeals to the semantics of geometric intuition more than that to that of the cultural background; in Fisher's case the report of the experiment means more a historiographic relation than with the dealing with reality (more science than technology), although experiments are about dealing with reality.

In the case of his reports, for instance the one published in the "Journal of Agricultural Science" CP15, the writing activity is not just

23 See [Canaparo, 2002]

a narration of the experiments or the quantitative results, but also a vehicle of the "communicable knowledge with which theories and interpretations are conducted (see [Canaparo, 2002, 100]):

> The crop records available at Rothamsted extend back for over 70 years. In Broadbalk wheat field 13 plots have been continuously under uniform treatment since 1852. [Fisher, 1972, 237]

Continuing with:

> From the series of observations it is possible to distinguish three types of variation in the wheat yield: 1) annual variation 2) steady diminution due to deterioration of the soil 3) slow changes other than steady diminution. [Fisher, 1972, 239]

Or:

> The average annual decrements when set out as percentages of the mean yield show a progressive advantage of the nitrogenous manuring. [Fisher, 1972, 246]

But also a theoretical proposition as the Theory of Polynomial Fitting exposed as the second part of the paper:

> If a quantity x have values $x_1, x_2, \ldots x_n$ at a number n of successive times, the general course of its changes may be represented by a polynomial,
>
> $$a + bt + ct^2 + \ldots + t^2$$
>
> in which t represents the time. [Fisher, 1972, 249]

But in his books, observations and experiments are separated: the first are known via description and not as a mathematical abstraction, which is more a visual literature not an abstraction or translation into another vocabulary, but it is the <descriptive statements> that work as <sinsign> of the real world. They stand for the existence of the individuals in a set, actually, they state directly that the set .

Analysing Fisher's texts one can find the interpretation of some images given in a real life experiment. As Calvino says:

> We could distinguish two kinds of imaginative processes: that one which comes from the word and arrives at the visible image and that one that comes from the

> visible image and arrives at the verbal expression. The first process turns for example a novel scene or a report of an event in the newspaper, and according to good or bad efficiency of the text we will bring the scene as if it was takeing place in front of us...[24]

And it is how in these literary terms, seen in the scientific literature, have become in Fisher a generator of images; images within the experimental and scientific world and the way of seeing things in that world.

Taking into account the extension of his method to other areas, one could consider Fisher as a positioning scientist, or self-positioning, in the <market> of other sciences, which he did well; coming from the agricultural laboratory to be a Galton Professor in London. This is totally different from the case of Blackett and his paratextuality. In this case he persuades a collectivity (of experimental workers, in his case) to follow a method, which would not only be successful in their field, but would also transform them into authorities in the interpretations of data.

As expressed in a letter to Darwin dated on the 13 of January 1939:

> For twenty years I have laboured, with more or less success, to get statisticians to appreciate the importance of Mendelian inheritance, and to get geneticists to appreciate statistical methods. [Bennett, 1982, 205]

In fact what he was doing was developing a sort of <science of the science>: a scientific method for an existing scientific method that was applied in other areas of knowledge. It is not the same to calculate the probability of some event as calculating the probability of the error in observing that event. What Fisher did, was precisely the essence of the

24 Possiamo distinguere due tippi di processi immaginativi: quello che parte dalla parola e arriva all'imagine visiva e quello che parfte dall'immagine visiva e arriva all'espressione verbale. Il primo processo giamo per esempio una scene di romanzo o il reportage d'un avvenimento sul giornale, e a seconda della maggiore o minore efficacia del testo siamo portati a vedere la scena come se si svolgesse davanti ai nostri occhi, o almeno frammenti e dettagli della scena che affioramo dall'indistinto [Calvino, 1993, 93]

so-called "Theory of Errors". In this sense, Gauss' notion of the Theory of Errors is in line with Fishers as both see them as the statistical treatment of observations, (see [Sheynin, 1971]).

This leads to a whole topography, a big space generated by his work, his images via the reports (in the form of papers) and methods on experiments and his parameters for the interpretation of the <real world>. This is the other point I will try to explain in this work, which is concerned with, what Canaparo calls, the <imachinery (imaquinación)> developed as a result of his <dealing with reality>. In [Canaparo, 2000], he explains the generation of visible planes as the manipulation of objects where the precision of the operations involve are the main drive in the construction of images.

Fisher is trying to give the reader an image. Fisher works on two levels of images: 1) the description of the experiment, which is totally visual, and 2) the images generated by the mathematical theory via geometric intuition or analogies with respect to the same object of the experiment. For example, referring to the explanation of the "plating method", he begins with:

> The process in general consists in making a suspension of a known mass of soil in a known volume of salt solution, and in diluting this suspension to a known degree. [Fisher, 1972, 374]

Or explaining the combinatorial method to achieve the test of significance:

> For 3 objects can be chosen out of 6 in only 20 ways, and therefore complete success in the test would be achieved without sensory discrimination, i.e. by "pure chance", in an average of 5 trials out of 100. [Fisher, 1935, 15]

A characteristic of his writing style is the explanation of concepts via examples. He does not develop algebraic work in his texts, nor does he propose mathematical assumptions, but he constructs the theory out of the example and proposes specific cases, and later, using to the power of mathematical language, he generalises them to any other applicable cases.

Fisher does not give the reader an image in every case, for example in his article Application of Vector Analysis to Geometry he does totally the opposite; although the title implies some geometrical intuition, his work is more concerned with the algebraic manipulation of terms, that is, focuses on the power of language (symbolic language, in our case):

> The torsion is evidently the rate of change of direction of r, that is
>
> $$\frac{\mathrm{r} \times \mathrm{r}'}{\mathrm{r}^2}$$
>
> but
>
> $$\mathrm{r}' = r' \times r'''$$
>
> therefore
>
> $$\mathrm{r} \times \mathrm{r}' = r'r' \bullet r'' \times r'''$$
>
> and so
>
> $$\mathrm{r}^2\mathrm{s} = r'r' \bullet r'' \times r'''$$
>
> where s is the vector measure of torsion. [Fisher, 1972, 61]

In the above algebraic process, the <narrative> goes from transformation to transformation in a way never <seen> or imagined by the reader, the only tools to follow are the use of the logic and the code of mathematical language justified in these cases.

In science the differentiation between a visual text and a non-visual one is clear; as Poincaré writes in his [1952] that the generation of images is given in a visual literature, in other words, outside of the pure algebraic syntax:

> To explain the nature of this hypothesis [talking about arbitrary hypothesis in physics], I may be allowed to use, instead of a mathematical formula, a crude but concrete image. [Poincaré, 1906, 197]

But in general this is not the case: Fisher in his texts is trying to use the observations to describe a reality external to it, just like a painter of the renaissance describes the interaction of the reality of an observer with its physity or circumstance. This is a radically opposite view compared with the one of the theoretical scientists that made out a theory first, that tried to explain in a logical and convincing way an a-priori argument that could be used in order to conclude.

Concerning the next step, we can say that the construction of the experiment is purely linguistic. As [Gross, 1996, vii-viii] says, rhetoric is more than window-dressing: it concerns the necessary and sufficient conditions for the creation of persuasive discourse in any field. Science cannot be excluded by fiat.

The problem is not to differentiate between mathematical language and colloquial language, but the language in general that melts in situations like Blackett's or Fisher's becomes a problem to the notion of knowledge[25].

Under this point of view we can generate a particular epistemology on observations and experiments:

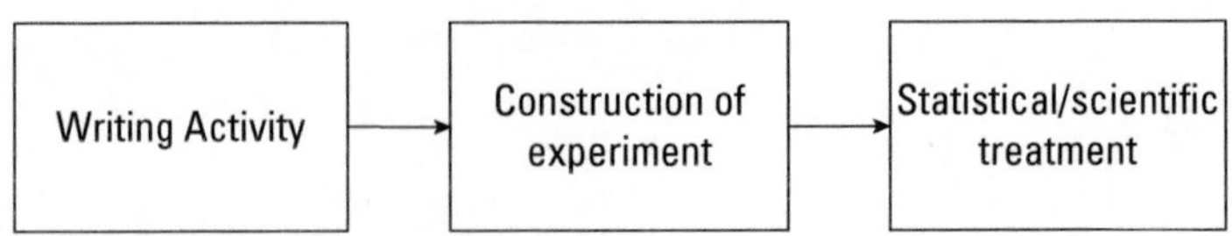

Experiment is a constructed concept, sometimes expressed in the language of the scientific community and at other times in the colloquial language that might position the work in further <markets>.

25 See [Canaparo, 2002, 102]

Chapter 5
Statistical methods for research workers: an epistemology not a methodology.

Introduction

In this section I will try to explain the generation of the breakthrough concepts in the statistical treatment of data which correspond to the next epistemic step: *generation of theory* all this in the context of the connection the steps corresponding to <Experimentation> and <Statistical Management>.

In 1925 Fisher published *Statistical Methods for Research Workers*; this is a book merely concerned with the statistical treatment of data, therefore in this section, the process moves from the experimentation which served as an "accurate" and "clean" source of data to the actual treatment of that data, and the generation of models to interpret and explain phenomena in nature as well.

Once the bulk data is generated, as members of a class, or elements of a set, then they are transformed into a mathematical entity, say $s=3.2$ and this is the form of a null hypothesis.

Instead of finding confirmation in experience as Blackett's case, Fisher tried to find confirmation in the statistical treatment. Of course, this involved conceptual structures rather than merely biological responses. This is a crucial step, because it is not in the observation or experimentation, but in the hypothesis and the theoretical knowledge that helps us conceptualise our experiential environment; i.e. it is the way to generalise our experience of the world.

As said by [Von Glassersfeld, 1995, 44] when commenting on Heisenberg: "the further natural scientists look into nature the more they realise that what they are seeing is a reflection of their own concepts" It is in this step when active writing begins.

But let's recapitulate the epistemic process: From the experiments a Fisher's *null hypothesis* was generated, this hypothesis is the main proposition related to the observations and it is the subject of scientific enquiry. This scientific enquiry is translated to statistical treatment and management of data.

His books are far from scientific papers, they are closer to technical works, and sometimes are closer to the real and everyday problems. This innovative point of view had not been written for the experimental workers hitherto: he converted the experimental worker into a statistician, as Statistical Science has and ends: "the reduction of data".

> In order to arrive at a distinct formulation of statistical problems, it is necessary to define the task which the statistician sets himself: briefly, and at its most concrete form, the object of statistical methods is the reduction of data.[Fisher, 1972, 278]

But what does this mean? It does not say reduction to the maximum which one would think could be zero data (or one), moreover, the process of reduction of data can be achieved in many forms. In Fisher there are three forms to do it, which he calls the three main problems of statistics:

> Problems of specification. These arise in the choice of the mathematical form of the population.
>
> Problems of estimation. These involve the choice of methods of calculating from a sample statistical derivates, or as we shall call them statistics, which are designed to estimate the values of the parameters of the hypothetical population.
>
> Problems of distribution. These include discussions of the distribution of statistics derived from samples, or in general any functions of quantities whose distribution is known [Fisher 1972, 280]

The way of dealing with these three problems is mirrored in the epistemology generated in this step.

The main mathematical contributions in the treatment of data are given in two concepts: 1) the *correlation coefficient* and 2) the *analysis of variance*. The first one refers to the explanation of causes and effects of events and the second to the treatment of errors in observation or errors produced by variables not controlled in the experiment, the latter is related with the more accurate, impartial and objective observations and with the refinement of the conclusions.

These two concepts interact with each other in the process of analysis and, at the end, generate a conclusion. It is important to have in mind that knowledge in Fisher's case is always related to experience, observations and experiments:

> The liberation of human intellect must, however, remain incomplete so long as it is free only to work out the consequences of a prescribed body of dogmatic data and is denied the access to unsuspected truths, which only direct observation can give. [Fisher, 1935, 10]

The word "only" in the above passage enhances what is a powerful critique of <handling> data, especially concerning statisticians. This is radically different from other scientists who did not have the conditions of "direct observation".

These scientists like Blackett, also wrote on the methods that should be performed to analyse phenomena, for example the suggestion of using some distributions, differential coefficients or inequalities.

Summarizing so far, there are changes in the process of constructing knowledge. They are ontological, hermeneutical and epistemological, but the most important part, in respect to the application of abstract theory, the generation of technical terms, and the use of active use language in a writing activity, begins in the abstraction from the experiment to the mathematical-scientific analysis of the evidence hitherto generated. As Maturana puts it,

> The fact that in a scientific explanation the phenomenon to be explained has to come out of the phenomenological domain different to that in which the generative mechanism -which would give birth as a result of its operation- takes place, constitutes to the phenomenon that has to be explained as a phenomenon in a phenomenological domain...[1]

And continues,

> Under this circumstances, the affirmation that scientists do with respect to the universal validity of the explanations and scientific declarations does not refer to a suspected revelation via them of an objective reality, independent, and therefore universal, but to its validity through the application of the operational coherences that are provoke in the opposite world or opposite worlds put in forward by the application of the criterion of validation which constitute them.[2]

This "validation criterion" or "criterio de validación" is reflected in the internal theoretic statistical area that is applied to the "objective reality" or "realidad objetiva".

This epistemic step begins in the representation of data. By 1921 Fisher had made the graphs of the actual yields of the plots of 67 years available in the Rothamsted database. The graphs showed progressive deterioration over the years, but changes in the mean yield; the mean yield rose between 1852 and 1860 and, after a low period in the 1870s reached again a maximum in the 1890s, then he began to see a pattern depending in the fertilizer applied that ranged from over 35 bushels per acre to 12. [Fisher, 1972, 242]

After the introductory chapter in his [1925] the first action was the construction of diagrams; in Fisher's own words:

> The preliminary examination of most data is facilitated by the use of diagrams. Diagrams prove nothing, but bring outstanding features readily to the eye; they are therefore no substitute for such critical tests as may be applied to the data, but are valuable in suggesting such tests, and in explaining the conclusions founded upon them. [Fisher, 1925, 27]

Divided epistemology.

Based on the first approach to the epistemological process in Fisher, with the incorporation of the statistical methods to the theoretical analysis of data, one can generate a second epistemology, divided in two cases:

1 Bajo estas circunstancias, la afirmación que hacen los científicos respecto a la validez universal de las explicaciones y declaraciones científicas no se refiere a una presunta revelación por medio de ellos de una realidad objetiva, independiente, y por ende universal, sino a su validez a través de la aplicación de las coherencias operacionales que ocasionan en el mundo o mundos puestos de manifiesto por la aplicación del criterio de validación que las constituye. [Maturana, 1995: 79]

2 See [Fisher J, 1978, 64]

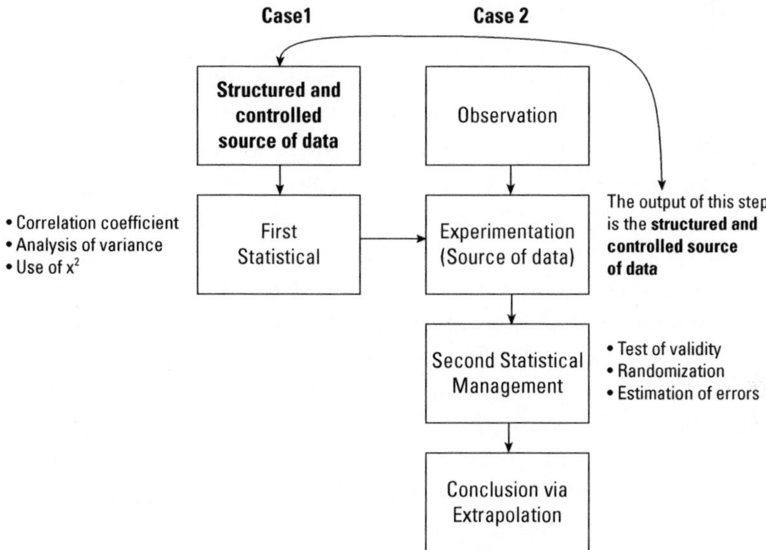

One case is when one has the control on the experiment and is the observer himself (as explained in the last section), and the other where the bulk data is given for further theoretical analysis by a third party.

For example if we have the data of the international prices of oil of the last 50 years and we have also, among other tables the economic growth of Japan. These observations were not done by the observer. The first action is to make graphs, plotting the values of each table. From a very quick observation one can find a relation between some of the variables, and then one measures the correlation and respective errors in observation, to establish a null hypothesis. This null hypothesis will serve to set an experiment to set a new aggregate of data that will be used to measure the validity and the "objectivity" of our conclusions.

Ontology of hypotheses.

In Blackett's situation hypotheses are real in the sense that they are able to find confirmation in the experiment. In Fisher's case, hypothesis are part of a measurable parameter derived from the relation observation-experiment, and their existence refers more to the logical existence; i.e. it exists if it does not come to contradictions in the mathematical sense. It would be natural to think in the opposite case, that it is because it exists in the objective reality that it does not lead to contradictions, but as seen in Fisher's construction of knowledge it is the other way around. Hypotheses are closer to the Kantian term "Heuristic fiction" (differentiating form the term "fact"), because it can be justified by the service it renders to science.

Only like this, hypothesis had the possibility of becoming theories or at least part of a theory that formed the body of knowledge that through further formalizations-confirmations became the explanation of reality or a model of it.

Summarizing, following [Von Glassersfeld, 1995], cognition serves the scientist's organization of the experiential world, not the discovery of an objective ontological reality. Knowledge is not passively received by the senses or by communication, but on the contrary, it is built up actively by the cognising subject. Another important factor, again, is the cultural environment.

Almost in the same year Fisher had expressed his views on statistics in the following words:

> The statistical procedure of the analysis of variance is essential to an understanding of the principles underlying modern methods of arranging field experiments [Fisher, 1925, 224]

And continues,

> The first requirement which governs all well-planned experiments is that the experiment should yield not only a comparison of different manures, treatments, varieties, etc., but also a means of testing the significance of such differences as are observed. [Fisher, 1925, 224]

Working in a laboratory such as Rothamsted, the problem was not the lack of information, but the non-processing of the information received, and furthermore, the ideas derived from that information. That could also be a reason why he concentrates on the method of analysis of the data and not in the problem of getting information.

Finally the following diagram illustrates the ontological difference from experimentation to the actual theoretical analysis.

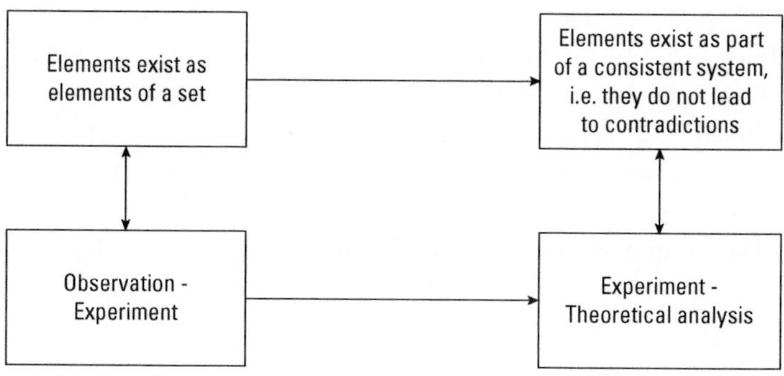

Hermeneutics of mathematical language.

Normally, after interacting with facts of reality, or reading any book, the reader- in his attempt to construct a relatively coherent model of knowing- collects a series of interpretations that make him see the world in different ways than other people. The situation in mathematics is different due to the mechanisms applied in the construction of a coherent cognitive model; and probably the power of algebraic and symbolic language is the key, for example in the hermeneutical sense one can reproduce a true replication of the concepts presented in a given work, the span of interpretations is reduced to a minimum. In other words, mathematical syntax makes in itself a coherent model.

More than qualifying mathematical language as universal due to his capacity of express and describe accurately facts of nature, we can say that it is universal because of its ability to diminish hermeneutical processes. When I say "x+3=5 then x=2", any person (trained in the area) can replicate without further interpretations (if it remains in the abstraction of the algebra of mathematics) what I am saying.

I am not saying that hermeneutics has no value in mathematics: one can interpret within the field of mathematics that the expression "x+3=5" represents a line in the Euclidean bi-dimensional space or a member of a family of lines, or a function with real values given certain conditions, etc. But of course this is far from interpreting a metaphor of the mystic discourse of a prophet in the a sacred book; this point helps to understand the subjective meaning and tries to compare it with the idea that words refer to things in themselves, i.e. words do not reflect things in nature as such, but they refer to abstractions from experience.

Correlation Coefficient

The first approach to the correlation coefficient is found in Galton's work. Francis Galton was studying the relation that existed between different characters in families, and observed that in a two-way table of measurements the contours of equal density in the table were elliptical; the narrowness of the ellipse about its major axis indicated the intensity of the relation between two characters. This property enabled to establish the correlation coefficient[3].

As pointed by Fisher:

> One of the earliest and most striking successes of the method of correlation was in the biometrical study of inheritance…it was possible by this method to demonstrate the existence of inheritance, and to "measure its intensity"; [Fisher, 1925, 139]

3 See [Fisher J, 1978, 73]

As seen in the diagram for the itinerary of documents (in chapter 4), the idea of correlation coefficient had been around in Fisher's papers since 1918, but reaching its major formalization 7 years later in [Fisher, 1925].

The correlation coefficient had been widely studied by W. S. Gosset, a chemist turned statistician due to the necessities of his work at the Guinness's Brewery, in Dublin. He had been student of Pearson in the Galton Laboratory in 1906 and 1907. Guinness's Brewery had large farming interests, especially in growing barley for beer; in consequence Gosset became involved in agricultural experimentation as well as with laboratory tests. However, Gosset and Fisher met in other circumstances, when Gosset visited the School of Agriculture at Cambridge and met F. J. M Stratton, the astronomer who happened to be Fisher's tutor in that time.[4] A paper in 1908 by Gosset treated the probable error of a correlation coefficient between independent variates and deeply influenced Fisher's point of view on the subject.

Fisher made a complete treatment of the correlation coefficient in [1925], dedicating the entire Chapter 6 to it. Again, his style is based on illustrations as he begins with an example of a table of data by Pearson and Lee, where it is recorded the stature of 1,376 fathers and daughters. And as noted by Fisher, The most obvious feature of the table is that the cases do not occur in which the father is very tall and the daughter very short, and vice-versa; this is the spirit of the correlation. In spite that the table is seen as slightly confusing, the plotting of values shows that they represent an elliptical form in which the major axis lies diagonally to the table. This geometrical configuration is noted by Fisher.

The definition of correlation coefficient is as follows:

$$r = \frac{S(xy)}{\sqrt{S(x^2) \cdot S(y^2)}} \quad \text{[Fisher, 1925, 147]}$$

Where x and y represent deviations from their respective means. This expression is derived by some statistical considerations:

4 See [Feinberg, 1980]

If x and y represent the deviations of the two varieties from their means, we calculate the three statistics s_1, s_2, r by the three equations

$$ns_1^2 = S(x^2)$$

$$ns_2^2 = S(y^2)$$

$$nrs_1s_2 = S(xy)$$

then s_1, s_2 are estimates of the standard deviations σ_1 and σ_2, and r is an estimate of the correlation ϱ. Such an estimate is called the correlation coefficient or the product moment correlation.

The above method of calculation might have been derived from the consideration that the correlation of the population is the geometric mean of the two regression coefficients; for our estimates of these two regressions would be

$$\frac{S(xy)}{S(x^2)} \text{ and } \frac{S(xy)}{S(y^2)}$$

so that it is in accordance with these estimates to take as our estimate of ϱ

$$r = \frac{S(xy)}{\sqrt{S(x^2) \cdot S(y^2)}}$$

Which is in fact the product moment correlation. [Fisher, 1925, 146-7]

Nevertheless, if one observes closely to this expression and assume that x and y are Euclidean n-dimensional vectors, the expression above is the inner product of the vectors, so geometrically what the correlation coefficient is doing is:

$$r = \frac{S(xy)}{\sqrt{S(x^2) \cdot S(y^2)}} = x \cdot y = \|x\| \cdot \|y\| \cdot \cos q$$

Being θ the angle between the two vectors. The results are corroborated in the geometrical approach: if r=0, the vectors are orthogonal, the same as the graphic of the data corresponding to the variables x and y. if it is 1, then the vectors are parallel and the correlation is perfect. (See [Plata, 2005]).

In fact the range interval for the cosine function cosine is $[-1,1]$ the same as the values for the correlation coefficient in statistical context.

In Blackett's case, the correlation is compared to the same geometric idea. When Blackett was referring to "break down the statistics", he meant to seek for the variables that could be meaningful for the problem to explain the phenomenon. And his method was to check "ceteris paribus" different variables of the problems and test them in differential coefficients, which can be interpreted as comparing slopes of vectors one to one, creating a parallelism with Fisher's case.

> If the variation of some yield with some one variable is being investigated, one must remember to look for and take into account as far as possible all the resulting changes in other variables [Blackett, 1948]

So what he did was to compute as many differential coefficients as he could, as expressed in his methodology:

> But the cultivation of method of thinking differentially about operational problems and of prediction by the variational method does seem of definite value.
>
> Thus, the first attack on any operational problem is often to estimate as many of the derivatives possible: first the tactical derivatives to judge what changes of tactics would lead to improved yields; then the material derivatives to estimate the effect of improved weapons [Blackett, 1948]

This means that for each entry of a row-vector in the data matrix M, one has to calculate the ratio between them. This is telling us that what Blackett was trying to find were parallel row or column vectors in the matrix, and if he looked for this, then the method was infallible!

This is better explained in the terms of *Linear* Algebra: Imagine the vector space $\Re$ and two vectors v, u in that space, then if v is linearly independent of u, then all $\Re^2$ can be spanned with this two vectors expressing any vector in the space as a linear combination of v and u. But if the two vectors are parallel, $v \parallel u$, then one is multiple scalar of the other one, geometrically it means that the only space that these two vector can generate is the subspace corresponding to the line that passes by the origin and is parallel to any of these two vectors.

But if we achieve this, we are in a great advantage in the 'breaking off the statistics', because we found that the ratio of any two variables is related in a proportional way, that is, one is a multiple of the other:

For instance for the above problem, one would assume the number of sightings to be proportional to the radar range when this is larger than the visual range, [Blackett, 1948].

Another way to connect this concept is to compute the slope *m* of the vector (if the vector $v = (a,b)$, then the slope is $m = \frac{b}{a}$); and precisely what Blackett states in his methodology is to compute the differential coefficient $\frac{b}{a}$, and this has a utmost value for the calculations which is precisely the equivalent to the correlation coefficient utilised by Fisher:

If we know that a and b are related (i.e. we found that the reduction to a 2-dimensional vector of the entries of a m-dimensional vector of the data matrix M are parallel), we are only 'moving' in a line and not in all the space of possibilities (spanning of the two linearly independent vectors), therefore our convergence to some hypothesis is almost sure, then

The calculation of the slope (i.e. the differential coefficient $\frac{b}{a}$, will tell us significant results: If we want the fraction to be zero or to be very large we only have to apply the correspondent limits to the respective variables of the reduced vector.

The difference between Blackett and Fisher in this respect is that Blackett takes the x and y as the data, while Fisher assumes that x and y are the deviation vectors of the data. What Fisher is measuring is the difference in the deviations, seen as vectors of a matrix of data.

Analysis of variance.

The arithmetical discussion by which the experiment is to be interpreted is known as the analysis of variance. [Fisher, 1935, 57] But what is the variance? To answer this question one must refer again to the *theory of errors* that in Fisher's terms is one of the oldest and most fruitful lines of statistical investigation: "The calculation of mean and probable errors show a deliberate attempt to find out something about that population".

This term is referred in Statistics as the mean value of the squares of these errors.

> The completion of the analysis of variance, when the yields are known, must be strictly in accordance with the structured imposed by the design of the experiment, and consists in the partition of a quantity known as the sum of squares (i.e. of deviations from the mean) into the same three parts as those into which we have already divided the degrees of freedom. [Fisher, 1935, 59]

Via the analysis of variance, we have the possibility of examining the variability of data in a population.

Analysis of variance is based on the idea that the total variance of the samples involved in the analysis equals to the sum of the sum of the variances of each sample.

$$Var_{total} = Var_1 + Var_2 + \ldots + Var_n$$

In the case where only two varieties are involved, then only two partial variances can be calculated and the total variance would seem very much to the Pythagoras' Theorem:

$$Var_{total} = Var_1 + Var_2$$

The above expression gives a geometrical approach to the problem, and in general allows to break complex problems into their components, and not only that, but from the point of view of the deviations of the problem, that served to identify possible errors of observation.

For the case of a 2×2 data matrix, the case of the analysis of variance is clear:

Let $x = (x_1, x_2)$ and $y = (y_1, y_2)$. x_i and y_i are the data of the table. When we calculate the mean of each set, that is $\bar{x}$ and $\bar{y}$ the resultant vectors lie in the subspace generated by the line if slope $m = 1$ that passes trough the origin. This vector is the most representative of all data generated in the sense of the most expected value. The variance for $\boldsymbol{x}$ is calculated as

$$s^2 = \frac{\sum (x_i - \bar{x})^2}{n - 1}$$

Now, let $\hat{x} = (\bar{x},\bar{x})$ and $\hat{y} = (\bar{y},\bar{y})$, then, the expression in the denominator is the square distance of the vector of data to the point of the expected value:

$$s^2 = \frac{\sum(x_i - \bar{x})^2}{n - 1} = \|x - \hat{x}\|^2$$

What Fisher is calculating is the average square distance of the deviations. Thus the standard deviation is exactly the distance of the deviations to the mean. And the total analysis of variance can be summarised in the decomposition of the total variance into the chosen block. This approach is useful to understand what happens with extended vectors that can move in more directions, not only two (n-dimensional space).

Analysis of variance is the generalization of the t-test (by Gosset) method by which one can test if the means of two populations are the same by comparing the mean of the two samples withdrawn from them.

In a similar way, analysis of variance also studies whether the mean of n populations are the same, by comparing their deviations not their means.

The example that Fisher works is an experiment with 5 varieties of crops divided in 8 blocks and the results of the yields of them are calculated.

What he wants to know is if the mean yield of each block is more or less the same in all the varieties. That is, to spot differences within the treatments for all the varieties.

Working by decomposing the variations of the treatments by the variation of the blocks, the conclusion is clear in the geometrical sense, because what one is looking is at the components of the vector which, by the way, depends on the mean components of the mean.

The difference in the effects can be given by many factors, even by chance. Thus if the variations are more or less the same size then different treatments have no real effect on the yields then the variations in the blocks are statistically irrelevant (i.e. they are by pure chance).

If the treatment does not make difference in the yields from the blocks, then the variation between blocks should be considerably bigger than the variation in the treatments.

Seeing this from a constructivist way, cognitive development in this stage is characterised by what Piaget [1970] calls *Equilibration majorante*, a process of equilibrium which goal is to eliminate perturbations when assimilating some experience; meaning by perturbations the deviations from an expected result. But it is precisely at this stage of equilibration that the learning process has a maximum. It is when a deviation from the expected result seeks for accommodation (re-establish equilibrium) when the cognitive process works.

Another interesting mathematical point not contemplated in his breakthroughs but valuable in the sense of geometric and topological interpretations in Fisher, is the use of dynamical systems in the explanation of genetic theory:

> In order to consider the chances in future generations we shall first calculate the appropriate frequencies for the case in which our gene is already represented in r individuals. In order to do this concisely we consider the mathematical function
>
> $$f(x) = p_0 + p_1 x + p_2 x^2 + \dots$$
>
> This function evidently increases with x from p0, when x=0, to unity when x=1...and in general the chance of leaving s genes will be the coefficient of xs in the expansion of $(f(x))^r$

And finishes with

> It follows that the total chance of leaving *s* in the third generation irrespective of the number of representatives in the second generation, will be the coefficient of xs in
>
> $$p_0(f(x))^0 + p_1(f(x))^1 + p_2(f(x))^2 + \dots$$
>
> Or, in fact, in
>
> $$f(f(x))$$
>
> This new function, which is the same function of f(x) as f(x) is of x, therefore takes the place of f(x) when we wish to consider the lapse, not of one but of two generations, and it will be evident that for three generations we have only to use $f\{f(f(x))\}$, and so on for many generations as required. [Fisher, 1930, 74]

The above explanation alludes to the concept of discrete dynamical systems, if the composition of functions represents iterations of the system in time. The time is measured in generations so the discontinuity is clear. In formal terms, the function is real valued and the Naturals act on the set of real numbers via the iterations of the function. This is what we call dynamical system.

It is important to note that the papers by [Cochran, 1980] and of [Somesh, 1980][5] on Fisher do not contemplate this, in our view important, geometric analysis.

Language and Literature in Statistics

Fisher is a writer that thought about his readers as active applicants, more than students of theory. His structure, in the case of the books, obeys to an organization that begins with simple concepts with no theory, or formal definitions or demonstrations, complicating the cases throughout the book. He bases his mathematical explanations in examples and a common factor is that he is always relating the theoretical concepts to objective reality.

Fisher states clearly the role of statistics, which may be regarded as i) the study of populations, ii) as the study of variations, iii) as the study of methods of reduction of data. [Fisher, 1925, 1]

Statistical Methods for research workers, his first book, is divided in eight parts, the largest part dedicated to the analysis of variance and, together with the section dedicated to the correlation coefficient, accounts for the 30% of the book pages.

The introduction- that is called Introductory- is more likely a chapter dedicated to the mathematical preliminaries, necessary to understand and 'read' the following chapters of the work. In part 4 of

5 See [Fisher, 1935, 40]

this *Introductory*, he explains what is the scope and purposes of the book.

> The prime object of this book is to put into the hands of research workers, and especially of biologists, the means of applying statistical tests accurately to numerical data accumulated in their own laboratories or available in the literature. [Fisher, 1925, 16]

It intends to be a sort of textbook, because he mentions the word student: "The book has been arranged so the student may take acquaintance with..." [Fisher, 1925, 18]

And he adds:

> The greater part of the book is occupied by numerical examples... there are no examples from astronomical statistics, in which important work has been done in recent years, few from social studies, and the biological applications are scattered unsystematically. [20]

The next chapter is dedicated to diagrams, which, in Fisher's words, prove nothing, but bring directly to the eye outstanding information. Technical work begins in chapter III, *Distributions*, where he explains, once again as a mathematical preliminary, the normal, distribution, Poisson series, Binomial distributions and χ^2. Once again, he refers to specific examples and tables of data to illustrate theory and applications of the distributions [Fisher, 1925, 46]

Indeed this book assumes that one should know about the formulae of the statistical distributions and the meaning of the parameters. But there are certain words and references that must be cleared out before continuing this work in order to understand better his achievements.

He is continuously talking about population, but what is a population?, it is a very basic concept for Fisher as stated above, it is the object of statistics, also population is referred not only to living or material individuals, but also the aggregate of results which is regarded as a population of measurements. This abstract population is what we know nowadays as the object of statistics, which we call <data>.

Fisher's case is important due to the generalisation that the abstraction allows, because he is not only studying the variations and errors on the estimations of the parameters related to the measurements of the observations on some physical (in the sense of a physical stable system) population, but he is also analysing the population itself through the abstractions of these empirical methods.

Based on observations and on the conciseness of statistics (as its aim is the reduction of data) he finds parameters (which he calls momenta: first momentum, second momentum, etc.) in order to infer some data to explain the behaviour and nature of those aggregates of individuals.

Fisher is more concerned with the finite cases than the abstract infinite cases which indicates his interest on the more practical and applicable knowledge (technology) than the theoretical (science) that do not deal with the physity of the world, although in some cases he attacks both.

The study of mean (first momentum of a "variate") as a parameter of variation on the behaviour of the population had been used widely, but the standard deviation had hitherto only been studied for less than a decade, Fisher considered standard deviation as a main parameter for the variation that must be taken into consideration together with the mean to study thoroughly the population. Mean and standard deviation are main tools for the analysis of experiments. Not only in the working papers of Blackett, but in Fisher's laboratory records, the basic analysis one can find is the calculation of means and variances, as illustrated in the next page, which is a record of Fisher's experiments on poultry during his stance in Rothamsted. [Fisher papers, Cambridge, not catalogued yet]

Studying the variance and standard deviation is concerning more on the <errors>, or deviations of the expected behaviour, than on the parameters that represent a population. Nevertheless, as seen above in the technical part the way to calculate deviations is linked to the parameters representing the population. Moreover, it was the study of variation that led to the concept of frequency distribution. The frequency distribution specified how frequent a variable "varied" in quantity, that is, in a number of observations. The number of observations of one type is called a variate. In our modern terms variate is a variable or more specifically a random variable.

-43 & -45. / 27 & 28.

Data relating to shank colour & barring.

Shank colour at hatching fell into 2 sharply marked classes – dark & light. With few exceptions the darks were barred & the lights non-barred. Though barred birds nearly always hatched with dark shanks the colour gradually disappeared, & barred birds when adult had either light or faintly tinged shanks. The whites were not always recorded.

Si. B	dk	30	cf. [illegible]
B	lt	2	
b	dk	1	
	lt	13	
G. B	dk	18	
B	lt	–	
b	dk	6	cf. [illegible]
	lt	17	
Wh.	dk	28	
	lt	31	
		156	

Constitution of both ♂♂ = BSi/b.l. Hence females produced =

B/	2	co. = 9.3 %
Bi	48	
b/	40	
bi	7	

Frequency distributions could be expressed in:

1. The proportion of the population for which the variate is less than any given value or
2. By the mathematical device of differentiating this function, i.e. the infinitesimal proportion of the population for which the variate falls within any infinitesimal element of its range.

So far, the problems, and above all, the kind of solutions and analysis performed by scientists and mathematicians in this field, included also the study of simultaneous variations of two or more variates. This led to the concept of correlation or as named by other writers, covariation. Fisher emphasizes on this and builds up a method for correlating one variable with another.

This first book was aimed to research workers, especially biologists, coaching them in the application of statistical tests to numerical data accumulated in their laboratories. In certain way Fisher is also a translator/creator, as when he is trying to explain and translate from one language (mathematical) to colloquial or other (biological) more familiar to biologists, in this way he created new concepts and new views. In a way, he is trying to ease the complexity of the problems so far posed and apply solutions in an actual way. As said by Fisher:

> The mathematical complexity of these problems has made it seem undesirable to do more than i) to indicate the kind of problem in question ii) to give numerical illustrations by which the sole process may be checked iii) to provide numerical tables by means of which the tests may be made without the evaluation of complicated algebraical expressions.[Fisher, 1925, 16]

In fact, this book quotes a lot of examples, "The greater part of the book" -Fisher says- "is occupied by numerical examples; and these perhaps could with advantage have been increased in number" [Fisher 1925, 19]

As said above, the visual aid was the one of the main tools for explaining the concepts, not only because "The preliminary examination of most data is facilitated by the use of diagrams" [Fisher 1925, 27], but also because the descriptive cases and explanations on examples.

In a way, the language and translations of Fisher and other scientists like Blackett try to express the external world as seen by a deep internal thought (in the constructivist view), and they the generation of a special way of seeing the world, a <*denkeweise*> a <way of thinking>.

In Fisher there is a balance, an equilibrium between the inner and the outer world that follows all his work, that if seen as a work of art gives a lot of elements of analysis normally hidden for the purely scientific reader. Images of the images are, as in the photographs in Operational Research or diagrams in Fisher's or data tables in both, the way of linking a problem of realism with a problem of abstraction, which is the scientific or even the artistic abstraction in the explanation of the world.

This duality generates terms related to notions of the whole activity of experimental science, because on the one hand it relates facts of experiments with reality and on the other the coherence of a system of propositions. This duality will act as part of the Peircean <ground>, giving the semantic connotation to the texts, so words as the following have determined meanings used in the context of the work of Fisher.

Random. This term refers to the random order that is a shorthand symbol for the full procedure of randomisation, by which the validity of the test of significance may be guaranteed against corruption by the causes of disturbance which have not been eliminated. [Fisher, 1935, 22-23]

Errors. This concept is associated with the error in observation that is derived from the cases not contemplated in the hypothesis. Error is not considered in the sense of comparison of the results with the reality, but basically in the first steps of the process (observation and experimentation).

A special meaning to the word is to refer to the acceptation of the null hypothesis 'when it is false' [Fisher, 1935, 20], but the term error is a functional part of the experimental science. This does not mean that errors should be done in every experiment, but in Fisher's words "it would be impossible to present an exhaustive list of such possible differences appropriate to any one kind of experiment, because the uncontrolled causes which may influence the result are always strictly innumerable. [Fisher, 1935, 21]

Equally probable. This term is referred to the principle of inverse probability that Fisher rejected; nevertheless, he used the term to assign objectivity in the process of randomisation. Equally probable only occurs in the design of the experiment when one tries to refine its accuracy and sensitiveness.

Experimenter. Experimenter is a part-statistician. He has to develop statistical analysis to validate his work.

Statistician. Statistician is a part-experimenter. He has to develop experimental analysis to validate his work.

Quantitative and qualitative. The quantitative properties of the hypothesis were translated from the qualitative characteristics of the observations. But in the process of mathematical abstraction, these quantities adopt a quality, like goodness, or validity, be translated again to the final interpretation/translation to the qualities of the objective reality.[6]

Validity. This concept is linked with the notion of the theoretical design of experiments and it has no connotations with the validity that could be generated from the comparison with reality. Validity is always in the internal process of the conduction of the experiment.

Cause. The cause is a factor of many as a result of a retrospective analysis on the events studies. Cause is related to the positive correlation of one variable and another. In the valid establishment of the relation lies the cause.

Terms like observation, experience, experiment and statistics are not included in this section because they are widely explained in the above sections due to their relevance in the understanding of Fisher's work; as seen there, they have also special connotations apart from the colloquial ones.

Syntactic view.

Within any discourse there is syntax; this is given by the structure of the sentences, the use of the verbs, tenses and narrative voice. Comparisons might lead to meanings due to the syntactic arrange given by them. For instance if one relates to facts seemingly totally disconnected like "the falling of the first leaf of autumn in London" and "the moment in which someone called Janet in Birmingham is caught irremediably by the rain" in the following way. "Janet was caught irremediably by the rain in Birmingham, and coincidentally at the same time the first leaf of autumn fell in London" probably will not mean much, in fact, this might be taken as a scientific statement in the discussion of simultaneity of facts in the context of, say, the relativity theory; but if I reverse the order of the sentence and say "At the same time that the first leaf of autumn fell in London, Janet was caught

6 The I occurs in philosophy through the fact the 'world is my world' (die Welt meine Welt ist). The philosophical I is not the man, not the human body or the human soul of which psychology treats, but the metaphysical subject, the limit of the world – not a part of it. [Wittgenstein 1933, 117-118, 5.641]

irremediably by the implacable rain in Birmingham", this last sentence can generate semantic interpretations that the first one cannot, and generate a metaphor full of melancholy and probably suggest the entrance of the autumn in Janet's life and the rain could be interpreted as a rain of tears when sadly and defenceless (as one caught in the rain) the cruel passage of time is present in her heart. In the above example one can observe that the order of sentences, the tenses of the verbs and the identification of the personal pronouns are crucial for the connotations of the entire discourse.

In a scientific paper or book, this is also present, sometimes without a theoretical background, but successful scientific texts have a recognizable structure that is always linked to the generation of knowledge.

In Fisher's case, the third person, which is identified with the experimenter, is the one who talks. The narrative identity is precisely the author as <experimenter in action>, and not as an <experimenter in writing>.

The way in which he constructs the propositions are from the latest to the first, for example:

> From 100 counts of bacteria in sugar refinery products the following values were obtained… It is evident that the observed series differs strongly from expectation [Fisher, 1925, 61]

The use of the third person is current in the construction of the theoretical part, as well as the use of the verb (observe). This is opposite to the use of verbs and persons in the other epistemic steps:

"Our experiment consists in mixing eight cups of tea" [Fisher, 1935, 13]. The experiment seems to be a common view, an objective truth undeniable to any common sense; the word "our" implies that it is us that are conducting the experiment. This type of discourse continues until he swaps and refers to the experimenter as a third person. And it is he who makes the observations:

> It is open to the experimenter to be more or less exacting in respect of the smallness of the probability he would require before he would be willing to admit that his observations have been demonstrated a positive result. [Fisher, 1935, 15]

Then he reverts again to "we". The sequencing of sentences is such that he refers first to the interpretations and then to the observations, which leads us to infer a constructivist point of view in his work.

"we" in this case should be taken as the constructor of knowledge. The first person pronoun, as Von Glasersfeld states, should be divided in two parts: the realist view, meaning the part of perceiving object and the constructivist view, where the first person constructs according to his experience. As Wittgenstein interprets it, we can say that the two-folded first person is one the one hand a cognitive subject and on the other has an undeniable biological dimension[7].

In the case of Blackett, syntactical analysis works in a different way, it is in the generalization of concepts that he applies the syntax, which is represented by mathematical and logical propositions; this is illustrated in the following diagram:

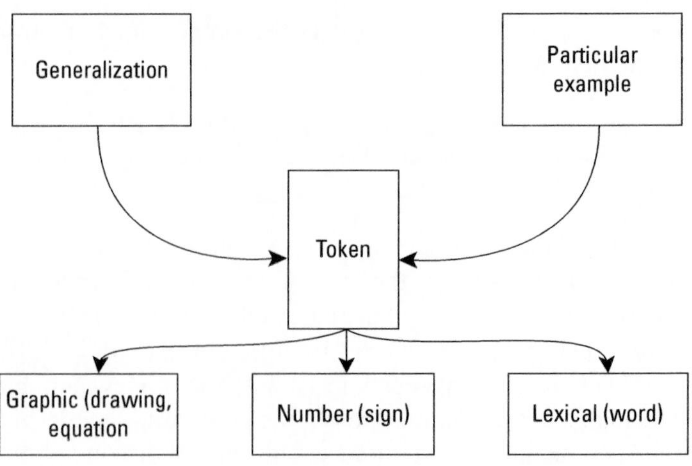

7 See [Von Glasersfeld, 1995, 62].

In this case, Blackett uses the concept of paratext, which is different in Fisher. The coincidence is that via language both achieve significative cognition. And the syntax worked from the idea of correcteness $p \Rightarrow q$ to the idea of meningfulness that relied on semantics, i.e "correlation with reality".

Semiotics.

The generation of signs evolves from the previous stage: the "statistic" in this case has a duality, it acts as a sign which refers to an object according to the system in which it exists, in a way this could be an <index>, nevertheless, it works also as an <argument> due to its relation with the parameters of population. Statistics are propositions for the characteristics of total population.

This is illustrated in the following diagram:

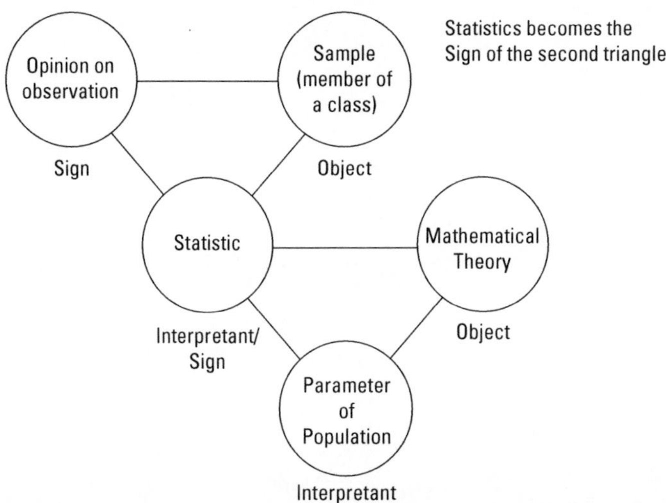

In the abstract level most of the signs generated are internal, and they are related either to the sign itself or to similar signs that work as objects for them.

The mathematical notation seen as signs and its manipulation within the symbolic language of mathematics are present in all the stage; in this sense signs like the variance are <symbols> because it is the convention that makes up the relationship between the sign and the object that it represents and not some representation of physical reality like in the two previous epistemic stages.

Chapter 6
Genetic theory of Natural Selection: A Parallel Case Study to business applications or Science in Action.

Introduction

The last steps in an executive (business) development process are the conclusions and applications. Both are relevant, one in the form of conclusive statements and the other emphasising on how the methods and arguments permeated to other sciences and areas of knowledge.

The extrapolation of the statistical argument, referring always to probability, is taken as the best approximation of truth; conclusions are not achieved in the same way as other cases. An opposite case is operational research where conclusions in form of recommendations or notes, were presented in the context of reports always comparing with other cases, for example this report issued by Blackett's section during the Second World War:

> In the ten months from August, 1940 to June 1941, the total weight of bombs dropped on the U.K. by the enemy was about 50,000 tons., that is at the average rate of 5,000 per month. The number of persons killed was 40,000 giving 0.8 killed per ton of bombs.
>
> The weight of bombs dropped by our offensive on Germany in the eight months from 22.6.41 to 15.2.42 was 15,000 tons, or at the rate of 2,000 tons per month.
>
> To calculate the casualties produced, one must allow for
>
> (a)The difference in type of bomb used
> (b)The difference in difficulty of target finding.
>
> (a)The enemy use mainly light case bombs with 50% charge. We use mainly G.P. bombs with about 30% charge.
>
> Experiments show that the blast effect of the latter is about one half that of the former, and the records of the Blitz show that the casualties and house damage are roughly proportional to blast.

And concludes,

> To reach the same relative scale of casualties, we would have, therefore, to multiply the effectiveness of our last year's offensive by a factor of 20.
>
> These arguments lead to the inescapable conclusion that our night bomber offensive has given negligible assistance to the Russians up to date. [Royal Society Archives, Blackett Papers, D64 D65]

The comparison of cases via the quantitative analysis of the elements is a main technique in Blackett.

In Fisher's case, conclusions are not always in the level of assertiveness required in an organisation like the Ministry of Defence, in Blackett's case the kind of action is always linked with the final conclusion of a report or a note. In Fisher's case, due to the nature of the process, the conclusions are in a more cautious form, for example:

> With such an arrangement, however, we have no guarantee that an estimate of the standard error derived from the discrepancies between parallel plots is really representative of the differences produced between the different treatments, consequently no such estimate of standard error can be trusted, and no test of significance is possible. [Fisher, 1925, 229]

This is following the idea of the rejection of the null hypothesis and not the acceptance of it as truth.

Another example refers to the expectation and future outcomes of an event.

> In the second season, however , there will be some further information, from the progenies of 25 plants each from those self-fertilised plants which happen to be double heterozygotes, and of these a certain proportion must be expected to be in coupling. [Fisher, 1935, 240]

The words *proportion* and *expected* are key to understand the kind of probable conclusions in Fisher.

But where does this conclusion come from?, the first step in the conclusion is to represent or re-present the elements seen in the experiment, this representation is not a description of something else, but a re-construction of the past experience from memory. In this case memory is not the natural human memory, but memory lies in the data generated in the experiment: It is important to note that it is not the more than 300 numbers in a data table, but the parameters deduced from them via statistics.

In this stage the writing activity is the central part. The communication of all the process is relevant as the conclusions are so. The relevance is measured in the connection with the objective reality, so the most important part is the interpretation of the experimental results, the translation of mathematical language to colloquial and the elements of language used to do it.

The objects of the three first epistemic steps shown in the present work (a) observation-experiment, b) experiment-theory, c) theory-conclusions) are assimilated as new material in each step, in the sense that assimilation is concerned with treating new material as an instance of something known. Following Piaget:

> No behaviour, even if it is new to the individual, constitutes an absolute beginning. It is always grafted onto previous schemes and therefore amounts to assimilate new elements to already constructed structures [Piaget 1967, 17]

Once is understood, the picture we get is quite different from the original one that the observer in the first instance grasped with his senses.

As said above the step from experiment to theory has the characteristic of the non-contradictory ontology. It is not in this step that experimenter is looking for practicality; therefore results are not measured by their practical value, but by their logical consistency.

This is concerned with the conceptual coherence as in the generation of theory, which is not present anymore in the last step, but evolves to an "ontological test" via corresponding the results with reality.

Experiments are phenomena always observable (for the experimenter), except when he is making mental operations; this theoretical abstract step is not observable. The reflection of this process has to be again related to the observables in the other steps.

This concept of equivalence is treated as a hermeneutic process. In a way, mathematics and especially statistics is the main tool with which the subject organises and manages the flow of experiences. It is important to know that referring to empiric experiences does not mean the empirical way of knowledge denying the operations of the mind reducing the cognitive process to a reception of objective sense of data. On the contrary, Design of Experiments, as we have seen throughout the present work, deals with the construction of what Von Glasersfeld call a relatively stable 'experiential reality' that is built without presupposing an independent world-in itself [Von Glasersfeld, 1995, 88]

Hermeneutics.

After representation is acquired, the concept of equivalence has to come into scene. Equivalence helps to differentiate the sameness of objects generated from the first steps to the following.

After all the statistical manipulation of data there should be equivalence between what was in the original observation and its representation in the conclusion.

To compare with the case of operational research, Blackett's hermeneutical approach is precisely the equivalence at every step of the work. In Fisher's case it is present only in this last part.

Re-presentation means the examination of a translation in the process of creating new knowledge. In this sense, the scientist is an interpreter, a translator and a decipherer, and the process is product of a successive interpretation.

Hermeneutical analysis in this case is explained by a successive process that might be divided in two: The idea of a unity that binds language and reality that implies that the world, or seeing it from the constructivist point of view, the conception of the world has only two possible interpretations: One of the memory (leading to ontological analysis), and the other hermeneutical via the successive interpretations. And it is in this process that the scientist maintains relation with the present.

As said in the last paragraph, the unity formed between language and reality is crucial in this step, because there are no more active or passive hermeneutics, the successive process maintains the ontological level in the writing activity. It is that from this point and on that interpretations and re-interpretations are only in the literature. Conclusions are never translated into actions in the real world, like in the case of Blackett. But the work of Fisher in applied mathematics and experimental science does not transcend to more spheres. For example, his interpretation of genetics is not applicable to an immediate action course, but lie as true in the scientific literature until further re-evaluation and re-interpretation. This succession kept the expressivity of the conclusions as the <narrative identity> is composed by the interpretation.

Following [Canaparo, 2001, 70] it is possible to distinguish reading from interpreting, due to the nature of perception in reading, interpreting is an epistemic step in itself, it is an hermeneutic construction that allows to glue the meaning of texts; but in this case the distinction lies in the consecutive re-presentations of the results, in any case, we can say that the hermeneutics of the process is achieved by another epistemic step called representation.

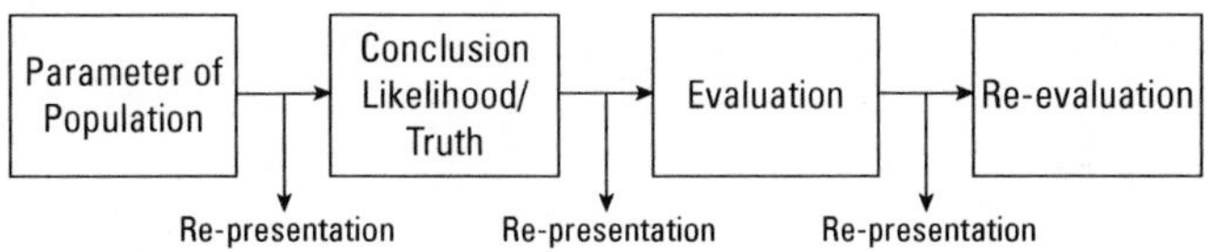

In a way this is a matter of transcending the theory to the empirical reality, and it is this way of translation from theory to conclusions, from technical language to normal language the crucial point, because without this dynamics the discourse would be incomplete.

Semiotics.

Semiotic analysis in this final stage refers also to the term <re-presentation>, and the relevant signs generated in the epistemic process are shown in the following diagram where the evolution from the previous stage in the parameter of a population derived from statistics leads to a new sign of truth based on scientific grounds.

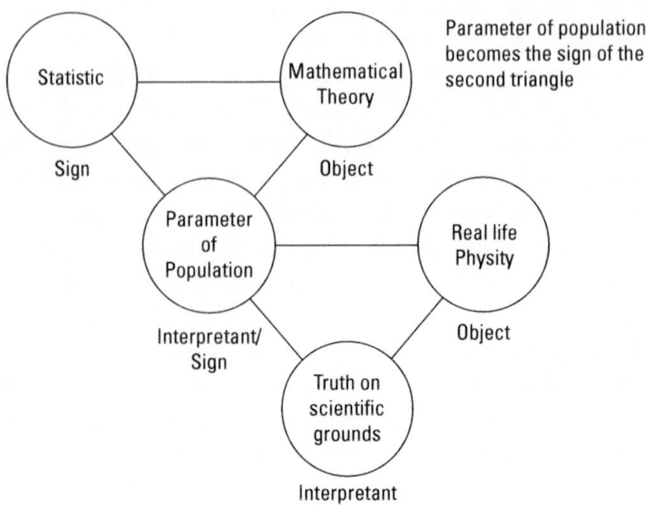

Ontology

Ontological activity moves in the level of <memory> In the sense of objects represented although not present at the moment and probably not even at one's attention.

An important fact is the difference of meaning within the language of mathematics that works as a dictionary: A dictionary only helps to expand the range of linguistic codes, when one looks for a word, the meaning is presented by using other words, but as Von Glasersfeld says [1995, 139], one cannot begin to learn from it (the dictionary) what the words encode. In Fisher's case the coding leads directly to the elements of sets always related to the visual utterance, but in this last epistemic step, words are explained by words. i.e the level of reality stays in the literary activity.

As in other fields of science, memory and the management of past information have its own dimension and size; in some cases apart from the biological dimension of the scientist, in this case, the past has an immediate presentation on the data tables, but this past does not exist outside these numbers and representations are achieved by the scientist/experimenter/observer himself. Following [Von Glasersfeld, 1995], "The ability to re-present objects to oneself is linked to language acquisition also in a very direct way". [60]

If there is a biological dimension outside of this records and writing activity, it is—as [Canaparo, 2000a] says—certainly strange to any communication and does not have possible writing. This is a direct consequence of rejecting the principle of inverse probability in the way Fisher did, and it explains why in order to manage scientific analysis, the translations and ontological interpretations of some objects of the "real world" had to be achieved first,

But the most relevant implication for this step is that also explains how the level of abstraction of the conclusions and results after the mathematical treatment do not transcend again to the objective reality.

Existence in this stage means existence in the memory; an element exists only if it is in the memory, i.e. if one is able to recall it then, it exists. This is the essence of the average man of Ortega y Gasset [1993]: one only recalls it when it is manageable for the human mind (not 300,000 data, but one representing all).

And it is in the form of written memory that one is able to recall it. That is why the connection of ontology and the level of written memory with hermeneutics and the successive re-presentations/ interpretations remain in the same plane.

Literature and Language.

In this step there are two factors that affect the narrative order a) the elimination of alternative interpretations of scientific data (represented in this case by the uniqueness of algebraic language) and b) the necessity from the scientists to demonstrate that the fabrication of his theory and/or explanations has been done correctly, or in short, that his method is valid.

This narrative order is probably the most relevant element in the construction of scientific facts.

The texts were relevant in the context of experimental scientists as a philosophy as understood by [Deleuze-Guattari, 1991], in the sense of drawing plans and inventing concepts. So the way in which this was achieved is consistent with the concepts of probability and the conclusions are more contemplative and explanatory than active (in the sense of taking action). Statistics seems to be the method by which one can conclude whatever the "owner" of the method wants. Therefore, truth is only a degree of likelihood based on the experience and inferred by the "scientific method". Even the counterargument of validity and representativity in Fisher's work had been surpassed.

Statistics was then the moral code to take decisions and this is a major heritage of scientific methods in industry and commerce, nowadays, in many companies and even in legal processes (such as the calculation of reserves related to the contingent liabilities in financial or insurance companies). Statistics as Fisher developed them represents a main argument to do it. Statistics for Fisher is a place from where to see, a method that includes the whole process, i.e. a space to have a vision, instead of just having a scientific perspective.

The kind of vocabulary used in the conclusions such as "not impossible" is clear that the degree of assurance is not 100%, normally a conclusion on the applications in Fisher's case would be like:

> It is thus not impossible that a mutant form, at first manifested equally by both sexes, should later under the action of selection, become confined to one sex only. [Fisher, 1930, 145]

But not only that, concepts like analysis of variance or correlation coefficient are expressed in the context of the conclusions in different way that in the context of the theoretical management.

> The effect of correlation between mates is ton haste, if the correlation is negative, or to retard if it is positive… such effects are not of importance, for even if the correlation were as high as 0.5, and mates had to be as much as parent and child usually are, the rate of decay would be more than halved. [Fisher, 1930, 4–5

Seen as an interpretation of the correlation coefficient and its applications the conclusions and explanations are not totally conclusive. The correlation between variables is not to be a quantitative one, but more qualitative the verbs hasten and retard are not measured in how much. Nonetheless there are cases that support in a 100% the diminution of the universe of possibilities; this is due to the assertiveness of mathematical theory and the interpolation of the semiotic, ontological and hermeneutical approaches in the writing activity:

> In a population breeding at random in which tow alternative genes of any factor, exist in the ratio p to q, the three genotypes will occur in the ration $p^2{:}2pq{:}q^2$. And thus ensure that their characteristics will be represented in fixed proportions of the population [Fisher, 1930, 9]

The binomial theory restrict the possibilities in the case above mentioned, nevertheless, in the conclusive part the proportions are not bounded by possibilities or number, but by qualitative properties like "lower" or "higher" without specifying the degree.

> The proportion must be even lower among lethals, unless indeed some of the obscure, though probably large, class of mutants which are lethal when heterozygous, be counted as dominant. [Fisher, 1930, 52]

The use of the words "probably" and "lower" are not determined in the general discourse, as we do not know the level but from the abstract level that something might be lower than some p or q given.

Again the inference process by which the results should be consistent with the theoretical treatment of the data is reflected only in a non-assertive way:

> One inference that may fairly be drawn from the foregoing considerations is that the widely observed fact that mutations are usually recessive should not lead us to assume that this is true of mutations of a beneficial or neutral character. [Fisher, 1930, 65]

The negation of a hypothesis is clear, but not its confirmation. The inference only says "may be drawn" and "should not lead" characterised by the non-assertiveness. Another case in which this type of conclusion is seen is in his text, when he "may assume that some genetic modifications are mutants" [Fisher, 1930, 59].

The theoretical deduction links the theoretical treatment with the observational experience, but never in a conclusive way:

> The theoretical deduction that the actual number of species is an important factor in determining the amount of variance which it displays, thus seems to be justified by such observations as are at present available. [Fisher, 1930, 98]

Concerning the applications one can find words as "believe" to refer to conclusions and forecasts.

> We have reason for believing that, with the evolution of new species, new mutations do sometimes commence to occur. [Fisher, 1930, 56]

The first person "we" is again in the scene. "we" is the observer, which in this case is associated with experimental scientists (see [54–55])

But one of the most non-assertive statements in the narrative of applications and conclusions is:

> On the other hand it does not seem, in the present state of knowledge, improbable that the greater part of the variance may be due to a cause special to heterozygotes. [Fisher, 1930, 63]

If theoretical language creates concepts by virtue of nomination, experimental science and applied mathematics create by virtue of predication. This is given in the same applications of the theory and in the level of explanation and relation to the physical reality. This is reflected in the conclusions and applications of the theory.

Syntax of language implies in this case syntax of reality. In the case of Blackett the mathematical symbols, together with reality, made a new structure, a new syntax in the construction of reality. In Fisher's case it is the relation between concepts of mathematical nature and objects of experimental reality that are meaningful and reason-able in the complete discourse, but above all in the conclusions and direct applications.

Under his view, the methods of statistics are essential in social studies. In fact, for Fisher the word statistics suggests a study of populations of human beings living in political union, and he always was very interested in the vital statistics and problems related to human beings.

> One of the earliest and most striking successes of the method of correlation was in the biometrical study of inheritance. At a time when nothing was known of the mechanism of inheritance, or of the structure of germinal material, it was possible by this method to demonstrate the existence of inheritance, and to "measure its intensity"; and this is an organism in which experimental breeding could not be practised, namely, Man [Fisher, 1925, 139]

But why to write on other fields? An explanation by the same Fisher is in a letter to Darlington dated January 9th 1936

> Dear Darlington,
>
> I am surprised and rather shocked, to hear that you should have experienced any difficulty in placing scientific papers. Although most of my stuff has been on subjects very different from yours, my own experience on this point may not be altogether irrelevant.
>
> When I started writing on mathematical statistics I supposed that a specialist journal was the most suitable place in which to publish. Biometrika was the only journal available. I published one paper there, which appeared in 1915. This was followed, in that and the following year, by two long editorial articles, under the names of a group of contributors, developing the solution I had given. The editor had not informed me that he thought any further development desirable, or invited me to co-operate, or, indeed, told me that he was doing anything about it. Next, he refused to publish a further paper giving new results and answering certain criticisms which he had embodied in the co-operative study. I was, forced to look elsewhere for the future, and published my answer in the Italian, or international, journal, Metron sending it directly to the editor to prevent its suppression by the nominated editorial agent of that journal in this country. Since then I have not offered any paper to Biometrika, and have published very little at all in journals specialising in mathematical statistics. In consequence, the methods I was developing appeared, usually a propos of some particular application, in something like 30 different journals.
>
> The only inconvenience I have felt in consequence of this is that, rather frequently, some mathematical writer, in search of proofs and of more comprehensive and coherent theoretical disquisition than he has come across, has published as new some result I have previously given, or, what is slightly more annoying, has asserted that I had given no proof of some important point, when he has merely overlooked it.
>
> Apart from this merely academic drawback, I am convinced that publication in non-specialist journals has been very much to my personal advantage, both in forcing me to think out problems from the point of view those likely to need

> their solutions, and in bringing my methods to the notice of a far wider group of workers likely to use them.
>
> The moral I am inclined to draw is that our scientific journals are, on the whole, too specialised for real utility that genetics, for example, has become quite unnecessarily isolated from, and unknown to, the larger body of zoologists, botanists and physicists, just because it was early provided with good specialist journals, so that the genetical discoveries, as they were made, only came to the knowledge of small group already interested in the subject. Consequently I say, on no account found a journal devoted to cytological genetics as many will, be inclined to advise.
>
> In England I should expect your papers to be welcomed by the journal of Genetics, Annals of Botany, Journal of Experimental Biology, Quarterly Journal of Microscopical Science, the Annals of Applied Biology, to name a short list. Anything directly, or indirectly, touching man would, of course, be welcome in the annals of eugenics, but I do not think I ought to take other genetical and cytological papers, apart from those which multiple factors and quantitative inheritance which are not taken care of elsewhere. [Darlington Papers, Bodleian Library Box 107 J47, J50].

This long quotation illustrates very well the circumstances in which Fisher began to specialise in other fields applying mathematics and extrapolating concepts.

But the must important fact in the literature at this point is the report of results via the extrapolation. This extrapolation is considered between the elements of the first epistemic steps and the theoretical management of data. In this case I will follow [Canaparo, 2002] and [Deleuze, 1985], on the ideas of the work of Henri Bergson of 1896 *Matière et Memoire*. These ideas refer directly to the cone of present past and future in which the pure becoming

There is a present moment in the time of writing. This moment works with a memory of the past and a *horizon d'attente*, waiting horizon, referring to all that is in the absolute present and has not possible writing. In other words, the text only can be structured from its narrative elaboration.

This present which is happening only can be named as happened: The past does not take place in the present which is not any more, it coexists with the present which has been[1]. And that what "was" is in fact the experiment itself. But the experiment itself as seen in the present work is closely related to images (visual literature). And it is because of that, that the explanation of Deleuze has a vital relevance in this part. The following diagram, taken from [Canaparo, 2000, 91], shows the cone of Bergson-Deleuze, but applied to our case:

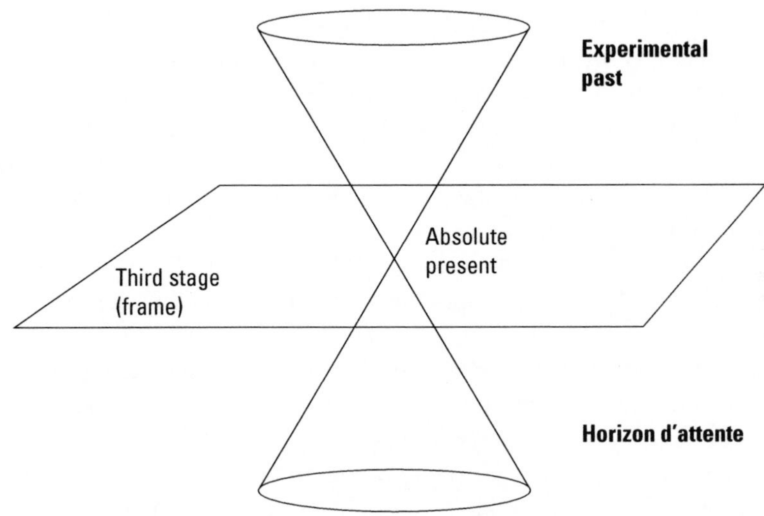

The present is moving from cone to cone, as the writing activity changes. This is reflected in the presentation of the conclusions like making sentences in which the length and clause-structure are reduced to a minimum point. This experiential past should be referred in Deleuzean terms to a present-past; a past referred to the present or in terms of present writing. Canaparo also proposes a present-future; which in this case is the spirit of the literature generated in the applications and conclusions of the experimental work in Fisher. Thus, this type of literature can be divided in actual, related to the objective reality and virtual, related to the theoretical work. In the conclusions we can see that there is a difference between perception and knowledge, attempt that Fisher tried to make clear since the beginning.

1 "Le passé ne succède pas au present qu'il n'est plus, il coexiste avec le present qu'il a été". [Deleuze, 1985, 106]

Historical and cultural approach.

Fisher and Blackett are characters that developed theories and achieved breakthroughs in science in their own fields. Probably being more famous for their work in Genetics or Physics (in fact in the data base of the British Historical Manuscripts one is catalogued as Geneticist and the other as physicist, the same as in references like *Who Was who*) but one thing in common was the development of applied mathematics in one case in the design of experiments and statistical treatment of data, and in the other leading to the development of operational research as a formal branch of mathematics.

They were contemporaries, although not from the same generation: Fisher's most important works are situated in the period of 1925 to 1935 and Blackett from 1930 to 1945. This was the time of the post-WWI and of WWII; both worked for government agencies, one for the ministry of agriculture (in Rothamsted Laboratory) and the other in the Ministry of Defence (in the Admiralty).

Their contact with the culture of that time is also reflected in their work. Marxism, existentialism, and even ideas in the air in the scientific artistic and cultural worlds of the time, for example the works of authors like Ortega y Gassett, Nietzsche, Poincaré, and Darwin, among others can be figured in some ways in their work.

The main breakthroughs in Fisher are the design of experiments, and concepts like *Analysis of Variance*, the use of the *chi square distribution* to analyse data, and the application of the *correlation coefficient* and *test of significance* in areas like Biology, Medicine, Genetics and Agriculture.

In Blackett's case, the breakthroughs can be considered in the insertion of scientists working in humanistic fields, like Economics, Production Management, and of course War Operations, also transforming the way of organizations in the design of reports and analysis of situations, e.g. with the militaries analysing some military situation, and the use of physical methods, like experimentation and the use of probability distributions to explain results and circumstances, and of course the born of Operational Research.

Ontology

The ontology of the elements change from epistemic step to epistemic step in the process, but definitely not in the physiological form. From observations to experiments the elements change their essence to become members of class, then to become arguments of a mathematical treatment and then to become part of a writing activity, in this case, the text is positioning a new ontology and a new paradigm of understanding.

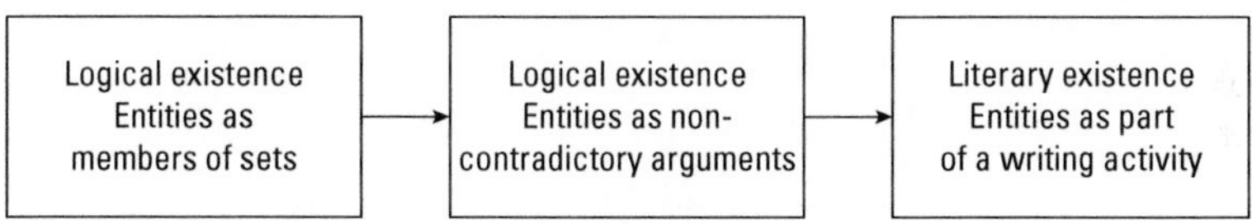

This required a new grasp of phenomena, of their relations and of the horizon of human possibility behind them.

Semiotics.

In the meantime that they interpret signs, they create them. Also the creation of signs changes and has dualities in each step. Their way of dealing with reality generates a complicated net of signs and representations. It is the relation between the structure of the reader and the structure of the writer that is the object of description in these sections. The following diagram illustrates the total set of transformation of signs in the epistemic process.

It is important to note that there are more signs generated in each step, but the only signs studied in the present work are the relevant in the construction of knowledge and for the explanation of the epistemology followed by Fisher and/or Blackett.

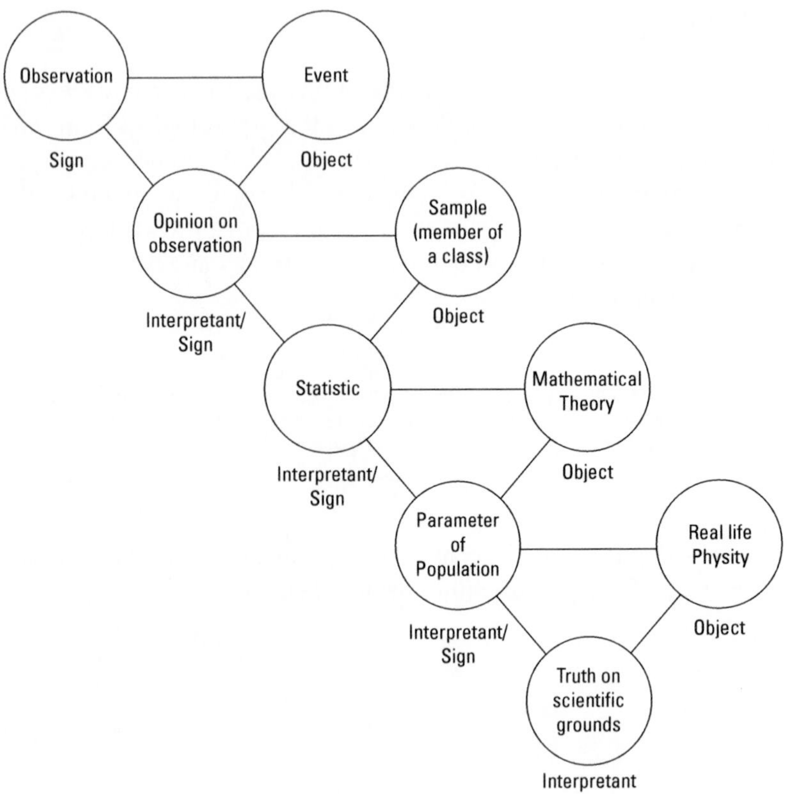

Literature and Language.

From a visual literature compared to a narrative of the laboratory report, to the extrapolation of arguments and the form of a text book to explain a method, the literary style and language changes from stage to stage. Syntax and semantics evolve in the process the same with the way of expressing the arguments and ideas. In Blackett's case the use of metaphors and similes are in constant use, but in Fisher the use of examples as a direct application is likely the case.

In the scientific discourse, weather of Blackett or Fisher, there is a formal and executive potential of words, of structure and of scripts to communicate. The language possesses the dynamics of fiction in the sense to invent and re-invent elements of the world. As Steiner says, Voiced truth is, ontologically and logically, 'true fiction', where the etymology of fiction directs us immediately to that of 'making'. [Steiner, 1989, 56]

Geometrical intuition is important in both cases; in one case Geometry functions as a special way of thinking, in fact is a way of thinking in images, in the other Geometry is a tool for semantic connotations. In both cases they reflect an art of thinking in images, based on them so the readers are led to have an actual "content" in the discourse.

The way in which literary elements evolve in Fisher's case is explained in the three sections: first coming from a visual literature compared to the description, depiction and construction of scenes, the next part regarded basically to the syntactic part, which is compared with the order of the elements like in the use of metaphors and the identification of narrative identities, and the third part with the construction of characters and concepts in the work, compared with different point of views in the literary world. The following diagram illustrates this transition:

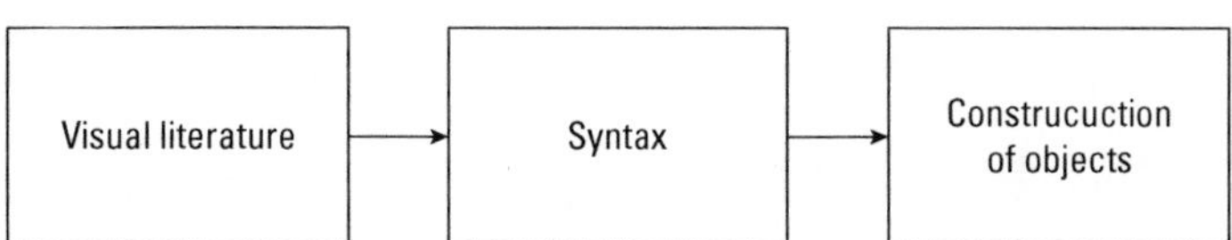

Epistemology

The two cases have similarities and parallelisms in the process of knowing and generating knowledge. The constructivist approach explains the case in all their stages from the objectivity of observations to the process of inference and synthesis and the writing steps of the end.

The process of knowledge is always linked to language; nevertheless other aspects of analysis have to be taken into consideration to explain the whole historiographical process like the biological dimension of the characters in their cultural context.

Although the present analysis sounds intricate, the case of applied mathematics deserves it due to its complexity. Throughout the present work the epistemological approach via radical constructivism and the relationship between historiograhic dimensions (including the institutions involved) and cultural environment (including political religious and personal views), lead to examinations in semiotics, ontology, hermeneutics, literature and language not diminishing of course the technical-mathematical part of the scientific breakthroughs, explained under some other interpretations (like geometric or analytic). Understanding the historical case as a whole might be clearer than other cases in which the case is seen under the specialization of one point of view.

Chapter 7
Kantorovich and Applied Mathematics: A Case Study in business development (cognitive rhetoric).

Introduction.

Kantorovich's case is a relevant one due its insertion in the corporate financial-economic fields. Kantorovich's scope goes from planning to any optimization process in what we might call today business development.

In his case the rhetoric between two radically different parts of the world (Capitalist-Communist regimes) was a barrier to break in the communication. The study of this case was of the utmost importance due to what I call a cognitive rhetoric, which implies, not only clear epistemological steps, but also a deep introspection on the way in which we (executives managers, consultants, or business developers) do our work.

Although Kantorovich wrote papers in a number of branches of mathematics, this work will concentrate on his work in applied mathematics, specifically in the area of economic management. However, his work on abstract mathematics, particularly on functional analysis, is closely related to the formulation of models used in his of applied mathematics work; for this reason, some of his pure mathematics work will also be, very briefly, referred to.

In 1975 Kantorovich was awarded Nobel Prize in Economics in 1975 for work on the optimal allocation of scarce resources; he shared the honour with Tjalling Koopmans. He used the mathematical technique of linear programming to advance the theory of optimum allocation of resources, as described in his 1959 work *The Best Use of Economic Resources*. The main ideas of this work were reproduced in the following year in *Mathematical Methods of Organising and Planning Production* that was published in 1960 in *Management Science*.

As stated by Koopmans, it was after a lecture on transportation models in 1949 given by Merrill M. Flood[1], that the mathematician Max Shiffman[2] pointed that he had seen similar ideas in Kantorovich's work on *The Translocation of Masses*[3] seven years before, in 1942. In 1952 and 1953, Flood himself in his paper made references to Kantorovich's work in his "On the Hitchcock Distribution Problem", a paper presented in the *Symposium on Linear Inequalities and Programming.*

The translocation of masses is a problem that analyses an important class of finite-dimensional extremal problems that are related to the analysis of certain questions of production organisation and planning. Although it is a problem linked directly with applied mathematics, or at least with its possible applications, the topic was included in Kantorovich's famous treatise on *Functional Analysis*, published originally in 1959. (see [Kantorovich, 1989, 225])

It was not until 1956 that Koopmans followed up the reference on Kantorovich and wrote him about the subject. Kantorovich replied and sent him some reprints, among them the one that was published in *Management Science*.

1 Merrill M. Flood served as the President of both The Institute of Management Sciences (TIMS) and for the Operations Research Society of America (ORSA). In fact, he was a founding member of TIMS and its second President in 1955. In 1961, he was elected President of ORSA, being the 10th to have the charge. He also served as Vice President of the Institute of Industrial Engineers from 1962 to 1965. Merrill provided the management science/operations research community with many decades of broad-gauged guidance. As early as 1936–1946, he applied innovative systems analysis to public problems and developed cost-benefit analysis in the civilian sector and cost effectiveness analysis in the military sector. Equally at home in his original field of the mathematics of matrices and in the pragmatic trenches of the industrial engineer, his research addressed an impressive array of operations research problems. His 1953 paper on the Hitchcock transportation problem is often cited, but he also published work on the travelling salesman problem, and an algorithm for solving the von Neumann hide and seek problem. Merrill Flood's career took him from Princeton University to the War Department, to the Rand Corporation, on to Columbia University and the University of Michigan. He was a pioneer and leader in demonstrating the applicability of his chosen field of operations research to problems drawn from all levels of society. For his many

In the set of reprints sent to Koopmans, there was also a paper written in 1949, on "The Application of Mathematical Methods in Problems of Freight Flow Analysis", published in the *Collection of Problems Concerned with Increasing the Effectiveness of Transport*, published by the Soviet Academy of Sciences (*Akademii Nauk SSSR*, Moscow-Leningrad, 1949), where he discusses transportation models for a single commodity as well as for many commodities (including empty vehicles), and a single-commodity model for a capacitated network, with applications to sections of the Russian railroad network.

In a later correspondence with Koopmans, Kantorovich confirmed that this paper was submitted at the end of 1940, but the publication was delayed due to war-time difficulties. To put this case in perspective with our other cases I should recall that the papers on Operational research in Britain by P.M.S. Blackett had widely circulated by then.

All problems considered in the 1949 and 1960 papers are what we now know as linear programming problems, a basic tool in Operational Research. For example, in technical books like [Ravindran, 1976], linear programming is treated precisely as a concern with the allocation of scarce resources.

contributions and the example he set for the profession, he was awarded ORSA's George E. Kimball Medal in 1983. His views of the profession are very well summed up by the advice he gave in his ORSA Presidential address in 1962 as follows. "First, cooperate liberally with all specialists who can contribute to the realization of new OR potentials, whether or not they work currently under the OR banner; and second, concentrate upon problem areas that are not only truly worthwhile, but that are also feasible in terms of research support and technological timeliness." (see [http://www.informs.org/History/Gallery/Presidents/TIMS/mflood.htm]).

2 Doctorate of the New York University in 1938, his Dissertation topic was on The Plateau Problem for Minimal Surfaces of Arbitrary Topological Structures being his supervisor Richard Courant

3 This work is considered a major breakthrough in pure mathematics as well as is reproduced in Sinai's compilation [Sinai, 2003] as communicated by S.L. Soboleff, Member of the Soviet Academy of Science under the "Compte Rendus (Soklady) de l'Academie des Science de l'URSS, 1942, Volume XXXVII number 7–8.

Acceptance of Kantorovich's work abroad, particularly at times of international East-West tension, was not smooth. However, leading specialists like Koopmans [1962], had viewed this matter with sober and generous views:

> If the effect of my introductory note is that Professor Kantorovich is now to be confronted with all the insights into linear programming theory, subsequently developed by several American authors, which he did not appear to have in 1939 –then my note has failed its purpose. But I so not see how or why this happened.[264]

And continues

> Neither do I understand the preoccupation of Charnes[4] and Cooper[5] with temporal priority. Is the glory of American developers of linear programming in any way diminished if it now turns out that, unknown to them, important aspects

4 Abraham Charnes (1917–1992). He was the seventh president of TIMS in 1960. B.A. in 1938, Ph.D. in mathematics in 1948 in the University of Illinois. He had a profound influence on scientific progress in arenas as diverse as the mathematics of operations research, optimization, statistics, fluid dynamics, as well as on functional areas of business including accounting, finance, human-resource planning, and marketing. Dr. Charnes was awarded the 1982 von Neumann Theory Prize of ORSA and TIMS (together with Cooper and Duffin). In September 1977, in an event held to honour his 60th birthday, he received the U.S. Navy Medal for Public Service, the Navy's highest civilian award. His contributions were recognized in many other ways. He was a fellow of ORSA, AAAS, and the Econometrics Society. Following service in the Navy during World War II, Dr. Charnes obtained a Ph.D. in mathematics at the University of Illinois. He then joined the faculty of Carnegie Tech in 1948. There his many accomplishments included pioneering work in mathematical optimization. His basic discovery of the association of linear independence with the extreme points of convex polyhedra was particularly noteworthy. He moved to Purdue in 1955 and to Northwestern University in 1957. At Northwestern he performed successful research in many disciplines, such as stochastic programming, generalized inverses, game theory, and nonlinear programming. Following his transfer to the University of Texas at Austin in 1968, he did seminal work together with W. Cooper and E. Rhodes that gave impetus to the new field of data envelopment analysis (DEA). A true pioneer in OR/MS, Dr. Charnes authored or co-authored over 400 articles, and seven books. (see [http://www.informs.org/History/Gallery/Presidents/TIMS/acharnes.htm])

> of linear programming models and theory had been anticipated in another language and another economic environment, with which communication has been somewhat difficult? If an element of national pride is involved, can we not justifiably point to a conspicuous discrepancy in the time span between development and application of linear programming ideas and techniques in the two environments? [265]

In addition, the equivalence between Kantorovich's method, linear programming (LP) theory and the algorithm for solution of "resolving multipliers" (also developed by Dantzig[6] under the more commonly known name "Simplex Method") is cleared out in [Koopmans, 1960]:

5 William W. Cooper was the first (founding) President of TIMS in 1954. In 1983, Professor Cooper was awarded the John Von Neumann Theory Prize by ORSA and TIMS. Dr. Cooper served on the faculty of Carnegie Mellon University from 1946 to 1976, where he was one of the founders of the Graduate School of Industrial Administration and the first (founding) Dean of the School of Urban & Public Affairs, now the H. J. Heinz III School of Public Policy and Management. He served as the University Professor of Management Science and Public Policy at Carnegie-Mellon before leaving for the Graduate School of Business at Harvard University. From 1976 to 1980, Dr. Cooper was the Arthur Lowes Dickinson Professor of Accounting at Harvard. In this role, he spearheaded the effort to reorganize the School's doctoral program. Author, co-author, or co-editor of 20 books and 450 scientific-professional articles, Dr. Cooper is a member of the Accounting Hall of Fame and a Fellow of the Econometric Society and the American Academy for the Advancement of Science as well as an honorary member of Omega Rho. (see [http://www.informs.org/History/Gallery/Presidents/TIMS/wcooper.htm])

6 In 1947 Dantzig made the contribution to mathematics for which he is most famous, the simplex method of optimisation. It grew out of his work with the U.S. Air Force where he became an expert on planning methods solved with desk calculators. In fact this was known as "programming", a military term that, at that time, referred to plans or schedules for training, logistical supply or deployment of men. Dantzig mechanised the planning process by introducing "programming in a linear structure", where "programming" has the military meaning explained above. The term "linear programming" was proposed by T J Koopmans during a visit Dantzig made to the RAND corporation in 1948 to discuss his ideas. Having discovered his algorithm, Dantzig made an early application to the problem of eating adequately at minimum cost. He describes this in his book Linear programming and extensions (1963): One of the first applications of the simplex algorithm was to the determination of an adequate

> Problem "C", [in Kantorovich's paper of 1960] while appearing still to have a somewhat special structure, is in fact equivalent to the general linear programming problem. [363]

In fact the proof was provided by Herbert Scarf[7] on the basis of a suggestion by Kantorovich himself, concluding that "Problem "C" can itself readily be put in linear programming form" [363–364]. Only to note, the multipliers of simplex method is named by Koopmans as 'shadow prices'.

And then Koopmans continues:

> At first sight it [the computational procedure] does not seem equivalent to Dantzig's simplex method, although is in a broader category with it in that it is also an iterative procedure in which trial vectors of quantities and of prices are successively revised in the light of profitability criteria. [364]

diet that was of least cost. In the fall of 1947, Jack Laderman of the Mathematical Tables Project of the National Bureau of Standards undertook, as a test of the newly proposed simplex method, the first large-scale computation in this field. It was a system with nine equations in seventy-seven unknowns. Using hand-operated desk calculators, approximately 120 man-days were required to obtain a solution. ... The particular problem solved was one which had been studied earlier by George Stigler (who later became a Nobel Laureate) who proposed a solution based on the substitution of certain foods by others which gave more nutrition per dollar. He then examined a "handful" of the possible 510 ways to combine the selected foods. He did not claim the solution to be the cheapest but gave his reasons for believing that the cost per annum could not be reduced by more than a few dollars. Indeed, it turned out that Stigler's solution (expressed in 1945 dollars) was only 24 cents higher than the true minimum per year $39.69. (see Article by: J J O'Connor and E F Robertson at http://www-gap.dcs.st-and.ac.uk/~history/Mathematicians/Dantzig_George.html

7 Prominent Yale economist who pioneered the use of numeric algorithms to facilitate the "computation" of equilibrium in general equilibrium systems. His approximation method to fixed points using simplicial subdivision was announced in 1967 and became the basis for famous 1973 monograph on Computation of Equlibrium, which launched the whole area of applied general equilibrium theory. Scarf also provided the first proof of the non-emptiness of the Edgeworth's "core" in 1967. He had previously provided in 1962, followed by joint work with Debreu in 1963, the core convergence theorem for a replicated economy. His counterexamples of stability of equilibrium (1960) helped, in several ways, to bury that research program. He is also renowned for

The rapid growth of mathematical economics and Operational Research during and after the Second World War stemmed from the same root: the application of mathematics to build and understand models that approximate special fields of human activity in “real life”. On the one hand, we have economists trying to solve economic problems based on some mathematical background, and on the other, we have scientists trying to extrapolate their methods in fields outside their proper scientific one. In other words, if mathematical models appeared in both cases, the root of their appearance and consequently their epistemic considerations are not the same.

The complexity and the interdisciplinary character of applied mathematics involving areas of optimization, inventory theory, and game theory included theories and methods from other areas of knowledge that at some point gathered in a single space, for example a physicist like Hitchcock in 1941, and an economist like Koopmans in 1951, independently developed what is considered the first useful optimization model: the transportation model.

Kantorovich (1939), a mathematician in the Russian central planning agency, developed several linear programming (LP) models for production and distribution including the trans-shipment model. Blackett (1939), a physicist in the Coastal Command of the Royal Air Force, developed the basis for what we know now as operational research in its interdisciplinary character. Much later, Dantzig (1951), a mathematician in the United States Air Force, developed the first generic linear programs and the simplex algorithm for solving them, which we have seen is equivalent to Kantorovich’s work, done nearly 15 years before. We see in this process, as so many times before in the history of science, a sequence of advances and rediscoveries.

his work on (S, s) inventory policy (1959). In his later work, Scarf tackled the problem of production sets with indivisibilities (i.e. non-convexities), a problem that has bedevilled production and equilibrium theory for a while—indivisibilities, after all, are the justification for technical increasing returns to scale. Resurrecting a result that he had found in 1963, Herbert Scarf (1986) noted that when such non-convex production sets are used, the non-emptiness of the core is not guaranteed. In order to handle non-convex production sets, Scarf has developed the method of "integral" activity analysis (e.g. Scarf, 1981, 1986, 1994).

Kantorovich and his contributions to applied mathematics.

Leonid Vitalevich Kantorovich studied mathematics at Leningrad State University, receiving his doctorate in mathematics in 1930 at the age of eighteen. From 1934 to 1960 he was a professor of mathematics at Leningrad University. He held the chair of mathematics and economics in the Siberian branch of the USSR Academy of Sciences (1961-1971), after that, he directed research at Moscow's Institute of National Economic Planning (1971-76).

Kantorovich's previous studies were basically in mathematics but he was involved in problems of the area of economics as the optimal distribution of scarce goods, a field to which he applied a variety of mathematical techniques.

He was one of the first to use linear programming as a tool in economics and this appeared in a publication *Mathematical methods of organising and planning production* which he published in 1939.

Kantorovich introduced many new concepts into the study of mathematical programming such as giving necessary and sufficient optimality conditions on the base of supporting hyperplanes at the solution point in the production space, the concept of primal-dual methods, the interpretation in economics of multipliers, and the column-generation method used in linear programming[8].

One of his most important works on economics was *The best use of economic resources,* which he wrote in 1942 but was not published until 1959. In this work Kantorovich applies optimisation techniques to a wide range of problems in economics.

He also proposed a theory to handle the economics of technological innovations. This had three components namely the effect on the producer, the effect on the consumer and, the novel part of

8 Kantorovich at the age of 25 produced a series of papers that contributed a new and fundamental orientation in functional analysis: the theory of semiordered spaces [Vucinich, 1984, 137].

the theory, the effect derived from the increasing economic potential arising from the innovation. This was mentioned in his Nobel Prize lecture.

Besides the application of mathematical methods, particularly mathematical programming, to economics, Kantorovich also worked in many other areas of mathematics. They include functional analysis and numerical analysis; within these topics he published papers on the theory of functions, the theory of complex variables, approximation theory (a topic in which he was particularly interested was the properties of Bernstein polynomials), the calculus of variations, methods of finding approximate solutions to partial differential equations, and descriptive set theory.

From 1929 he worked on the theory of analytic sets and the Baire classification of functions. This work continued through the early 1930s then in the late 1930s he studied ordered topological vector spaces. Later in his career he also became interested in the theory of computer architecture. His remarkable contributions to mathematics, economics and computers were published in over 300 papers and books.

Among many distinctions and honours he obtained: the Order Signe of Honor (1944), the Order of Labour Red Banner (1949), was elected Member-Correspondent (1958) of the USSR Academy of Sciences and later Academician (1964), Lenin Prize (together with V.V. Novogilov and V.S. Nemchinov in 1965) Order of Lenin (1967), he also became a fellow of the Hungarian Academy of Sciences (1967), the American Academy of Arts and Sciences (1967) and a Boston Fellow of the Econometric Society (1969)[9].

9 For more on his biographical notes, see [Sinai, 2003, 549–550], who reprinted an article of J. J. O'Connor and E. F. Robertson based on the biography of Kantorovich published in *Encyclopaedia Britannica*. Another source is Roy Gardner's paper of 1990 *L. V. Kantorovich: The Price Implications of Optimal Planning*, published in the *Journal of Economic Literature*. Also see *Nobel Lectures, Economics 1969–1980*, Editor Assar Lindbeck, World Scientific Publishing Co., Singapore, 1992

A brief biography

Kantorovich was born in St. Petersburg (Leningrad) on 19th January 1912. His father, Vitalij Kantorovich, who was a doctor, died in 1922 and it was his mother, Paulina, who brought him up.

According to his autobiography which was written at the time of the Nobel Prize award (1975) and later published in the book series *Les Prix Nobel/Nobel Lectures*[10]., his first interest in sciences and the first displays of self-dependent thinking manifested themselves about 1920. He entered the Mathematical Department of the Leningrad University in 1926, and thanks to the lectures of E. Tarle, he was not only interested in mathematics but also in political economy.

At University, he attended lectures and worked in seminars of V. I. Smirnov, G. M. Fichtengolz, B. N. Delaunay; and worked with I. P. Natanson, S. L. Sobolev, S. G. Michlin, D. K. and V. N. Faddeev."

The Petersburg mathematical school combined theoretical and applied research, and when he graduated from university in 1930, he started his research in applied problems. The speech of industrialization of the country created an appropriate atmosphere for such developments. In 1936 he worked at the Leningrad University and in the Institute of Industrial Construction Engineering."

At this time, the development of functional analysis was large and became one of the fundamental parts of modern mathematics. Kantorovich's studies were focused in the systematic study of functional spaces with an ordering defined for some of pairs of elements. And as said in his own words:

> This theory of partially-ordered spaces turned out to be very fruitful and was being developed at approximately the same time in the USA, Japan and the Netherlands. On this subject I contacted J. von Neumann, G. Birkhoff, A. W. Tucker, M. Frechet and other mathematicians whom I met at the Moscow Topological Congress (1935). One of my memoirs on functional equations was published as a result of the invitation extended to me by T. Carleman in Acta

10 From Nobel Lectures, Economics 1969–1980, Editor Assar Lindbeck, World Scientific Publishing Co., Singapore, 1992

> Mathematica. Functional Analysis in Semiordered Spaces, the first complete book of our contributions in this field, was published in 1950 by my colleagues, B. Z. Vulikh and A. G. Pinsker, and myself." [Lindbeck, 1992]

His theoretical and applied research had nothing in common. But later, especially in the post-war period, he succeeded in linking them and showing broad possibilities for using the ideas of functional analysis in numerical mathematics. This was proved in his paper, *Functional Analysis and Applied Mathematics*, which seemed, at that time, paradoxical. In 1949, the work was awarded the State Prize and later was included in the book, *Functional Analysis in Normed Spaces*, written with G. P. Akilov in 1959.

During the dacade of 1930 he began his first work on economics. In 1938, he was not only professor at the university, but also he was a consultant for the Laboratory of the Plywood Trust solving a very special extreme problem. Economically speaking, it was a problem of distributing some initial raw materials in order to maximize equipment productivity under certain restrictions. Mathematically, it was a problem of maximizing a linear function on a convex space.

This particular problem turned out to be very typical. He found many different economic problems with the same mathematical form: work distribution for equipment, the best use of planting area, rational material cutting, use of complex resources, distribution of transport flows. This was reason enough to find a general efficient method for solving the problem. The method was found under the influence of ideas of functional analysis which he named the "method of resolving multipliers".

In 1939, the Leningrad University Press printed his work called *The Mathematical Method of Production Planning and Organization* which included the formulation of the basic economic problems of this kind (resources allocation), their mathematical form, a sketch of the solution method, and a first discussion of its economic sense.

In essence, it contained the main ideas of the theories and algorithms of linear programming. The work remained unknown for many years to Western scholars. Later, Tjalling Koopmans and George Dantzing, found these results independently. But their contributions remained unknown to Kantorovich until the middle of the 50s.

Kantorovich's work on economics took two main paths: 1) The methods for solving extremal problems and their generalization in the sense of the application to other areas of industrial and economic activities, moreover, the extension of the method and analysis to planning problems, and 2) A mathematical model for these problems such as, non-linear problems in functional spaces and the applications of mathematics, mechanics and technical sciences.

Kantorovich interrupted his studies by the time of the war and during the war he worked also as Professor of the Higher School for Naval Engineers. When he returned to Leningrad in 1944, he worked at the University and at the Mathematical Institute of the USSR heading the Department of Approximate Methods. At that time, he became interested in computation problems, with some results in the automation of programming and in computer construction.

He worked in 1948-1950 on the problem of the Leningrad Carriage-Building Works. Here the optimal use of steel sheets was calculated by linear programming methods. His book of 1951 summarized his experience and gave a systematic explanation of our algorithms including the combination of linear programming with the idea of dynamic programming (independently of R. Bellman).

In the middle of the 50s, the interest in the improvement of economic control in the USSR increased significantly, and conditions for studies in the use of mathematical methods and computers for general problems of economics and planning became more favourable; precisely at that time, he contacted foreign scholars in this field. As a particular result, thanks to the initiative of Tjalling Koopmans, the 1939 work was published in *Management Science*.

The field attracted a number of young talented scientists, and the preparation of such hybrid specialists (mathematician-economist) began in Leningrad, Moscow, and some other cities.

Kantorovich was elected Corresponding-Member of the Academy of Science in 1958 and came to Novosibirsk in 1960. Out of his group in Novosibirsk, a number of talented mathematicians and economists emerged."

Nobel Prize Lecture

One of the most interesting parts of Kantorovich's view is reflected in the Nobel price lecture, where he states among other things the importance of functional analysis in applied mathematics, and puts forward the importance of concepts like control and planning in the contexts of a controlled economy like the Soviet economy of that time. He also points out some other connections he considered of importance to develop his work in this field.

It is important to note the existence of a form of paratextual activity, since the lecture was done in 1975 (although reprinted in *The American Economic Review* in 1989) he referred to achievements and ideas developed at the end of the 1930s, nevertheless this *paratextuality* differs from others' in the sense of the commercial impact that the original work might have had. In fact the reception of his ideas in Western circles was in some way the work of Koopmans, who defended his work in many ways.

> In our time mathematics has penetrated into economics so solidly, widely and variously, and the chosen theme is connected with such a variety of facts and problems that it brings us to cite the words of Kozma Prutkov which are very popular in our country: "One can not embrace the unembraceable". [Kantorovich, 1989, 18]

In our time, means clearly 1975, the lecture is a current lecture in which the state of the art of the discipline was the main topic of the occasion.

> I want to restrict my theme ... mainly to optimization models and their use in the control of the economy for the purpose of the best use of resources for gaining the best results. I shall touch mainly on the problems and experience of a planned economy, especially of the Soviet economy. [Kantorovich, 1989, 18]

The fact that the centralised economy in the Soviet Union had appear as a context to develop methods of organising vast amount of means of production and resources, might have had a crucial impact for the presentation of the theory. As Kantorovich points out in his lecture,

> The following problems are related both to the economic theory and to the practice of planning and control. [Kantorovich, 1989, 18]

It is clear that he did not have to make explanations of the phenomena or actions or the underlying philosophy or system of economy, but a clear objective: "practice of planning and control". In this sense a main concept in his work was "planning", which in this case was not referred to future actions based in a economic theory only, but to a numerical activity, in which the quantification of variables was of the utmost importance to communicate problems.

> This planning must be so detailed as to include specific tasks to individual enterprises for specific periods and to that common consistency of the whole this giant set of decisions was guaranteed. [Kantorovich, 1989, 19]

Thus, planning became a problem in itself, as it was not an isolated variable, but required a "consistency" in the system, and therefore a specific planning, in other words the planning of the planning, which was achieved more easily via mathematical tools, as he says:

> It is clear that a planning problem of such scale did appear for the first time, so its solution could not be based on the existing experience and economic theory. [Kantorovich, 1989, 19]

In this sense, economic theory was, probably for the first time, altered in its essence, as the necessity of a form of study in which a systematic control of the economic processes and entities was achieved. In this sense, mathematical language was a very convenient tool, not only to solve problems related to numerical analysis, or decision making, but also to communicate in the best way the more suitable further actions to be taken, subjected to some constraints. In this view of economics it is needed "the proper information and methodology to provide decisions that are in accordance with general goals and interests of the national economy".

It is important to note that it is not the aim of the discipline the explanation of an economic theory; instead, the economic discipline serves as a basis for an "operational research" approach to economic matters, especially in an economy where the "invisible hand" of Adam

Smith which moved the forces of market, did not apply, for example, "it was unclear and open to discussion whether a land rent should exist in a society where land is in the possession of the people or whether such an index as the interest rate has a right to exist".

Planning and Control: an inseparable duet.

Under this view, we can point out that another important issue is the fact that in Kantorovich's discourse planning is always related to control as an objective of economy.

> The decisions of different control levels and from different places must be linked by material balance relations and should follow the main object of the economy. [Kantorovich, 1989, 18]

The control problem was then to build a system of information, (accounting, and economic indices) that allow local agents whose main task was decision-making, to evaluate the advantage of their decisions from the point of view of the whole economy, in other words to put a quantitative system which measured the performance of certain policy of action, in Kantorovich's terms, to have the possibility to "check the validity of the work of local organs' activity also from the point of view of the whole economy".

New problems of control of the economy and new methods put forward the question of the most efficient structural forms of control organization; which became the true underlying problem of what we might call a basic tool of operational research: linear programming. The aim was the perfection of the control system as well as changes in the economy itself; inevitably, connections with the increase of its scale and the complexity of links with new problems and conditions arose.

The problem of the most efficient structure of a planning system has also a scientific aspect, but its solution is not well advanced. In this way, a solution was via numerical comparisons,

> It [the solution] must contribute not only general qualitative recommendations but also concrete quantitative and sufficiently precise accounting methods which could provide the objective choice of economic decisions. [Kantorovich, 1989, 18]

That will help to evaluate further economic decisions. In this way the duet planning control, is inseparable of the problems that were inherited to what we know now as operational research.

This complexity, led to gather problems of different roots into a single arena:

The problems.

One of the first epistemic steps in linear programming processes is the problematization, i.e. how is the problem posed in order to lead to a meaningful solution. In some cases like in [Kantorovich, 1940] and [1942] the problem is given in a specific way, trying to measure exactly the performance of some variables, in other cases the problem is given as a precise mathematical problem, for example in [Dantzig, 1963, 32]. Historically, a number of theories have developed in the way they did, mainly due to the way in which the basic problems to solve were presented.

This discussion begins in the problematisation as it tackles the problem of analysis concerning the efficiency of the methods used, therefore one must put the highest attention to the epistemology of the process, more than merely comparing classical parameters like the error rate or computing time, which are only a partial steps of the whole process. In this study I attempt to explain the epistemic and knowledge problems of the area known under the name of Operational Research,[11] and try to identify some of the foundations elements that pertain to this area of knowledge.

11 In this case, Operational Research refers to Linear programming, statistical methods and other applications of mathematics to problems of real life.

In the work of Kantorovich, it is important to define precisely the relationship between the objects of the real world and its perception and their role in the generation of knowledge.

The problems Kantorovich tackled in those times were complex problems of economic control that were generated by the contemporary development of the economy, i.e. by the so-called scientific-technical revolution. Problems like the prediction and control in different branches of the economy, or the control in the rapid changes in production and technology within the national economic scheme were on top of the list, however, problems of estimating technical innovations and the general effect of technical progress, as well as problems of ecology connected with the deep changes of the natural environment under the influence of human activity were considered. But one of the important problems to notice was, in the words of Kantorovich, "the prediction of social changes and their influence on the economy".

Of course, in the capitalist economy this last problem is important, but in the socialist economy of that time it represented an ideological problem as well, and given that Marxist theory pre-conformed the methodological background for the control system and that there have never before had a practical use of Marx's ideas "there existed neither experience nor sufficient theoretical foundation for the solving of these hard problems".

As pointed by Kantorovich, these problems used to be solved, on a practical basis, by governmental bodies during the first years of the then newly born state:

> Nevertheless the problem of building up an effective economic mechanism was resolved. I have no possibilities to describe it in detail but I just wish to point out that the system of planning organs was created on the initiative of the founder of our state V. Lenin and simultaneously on the same initiative a system of economic accounting (hozraschet) was introduced which gave a certain financial form of balance and control of separate economic activities. [Kantorovich, 1989, 19]

During and after the Second World War, industrialisation systems forced new ways to control the economic mechanisms, problems of incompleteness of many production or transportation by train arose immediately; "this led to the natural idea to introduce and use quantitative mathematical methods".

And although there were mathematical methods used in economy, like the demand models of E. Slutsky and A. Konjus, the growth models of G. Feldman, the balance analysis done in the *Central Statistical Department* of the USSR, which was later developed both mathematically and economically using the data of the US economy by W. Leontiev and more, the main problem was about optimization models which only appeared in the scene in the late 1930 (and later independently in United States).

The optimising approach is here a matter of prime importance. The first approach is the treatment of the economy as a single system[12]; this can be seen in the technical analysis made by Isbell and Marlow in their paper of 1961 "On an Industrial Programming Problem of Kantorovich", where they state the problem:

> We suppose that there are available N technological methods of production $P_s(s = 1, \ldots, N)$. Each method is completely described by $l + m$ real numbers $x_1^s, \ldots, x_l^s, y_1^s, \ldots, y_m^s$. The first l numbers refer to the l types of desired final products, the remaining m numbers to other materials (factors of production or intermediate products). [Isbell, 1961, 13]

The problematisation contemplates not only methods of production, different intermediate and final products but also factors of production, which if we assume linear then we might express the optimising system as the linear combination of the variables:

> The total production in each type is given by the linear expressions
>
> $$\sum_{s=1}^{N} p_s x_i^s, \sum_{s=1}^{N} p_s y_j^s \text{. [Isbell, 1961, 14]}$$

12 In fact, an economic system in relation to the optimisation process is outlined by Kantorovich in the following terms: "this well-known model [optimisation model] which is based on the description of an economy as a set of main kinds of production (or activities,—the term of professor T. Koopmans), each characterized by use and production of goods and resources". And continues, "It is well-known that the choice of optimal program i.e. of the set of intensities of these activities under some resource and plan restriction gives us a problem to maximize a linear function of many variables satisfying some linear restrictions". [Kantorovich, 1989, 19]

Control allowed the efficient systematization of enormous information material, which became later, the core of decision-making.

The aim of control led to the establishment of a optimisation model, which ensured the correct and most profitable use of resources.

On the 29th September 1942 Kantorovich wrote a note entitled "On the Translocation of Masses" communicated to the academy of science of the USSR by S. L. Soboleff, in this paper he states the study of what is known as the Monge-Kantorovich problem or the "mass transfer problem", which consists basically in minimizing the functional:

$$c(\mathrm{m}) := \int_{X \times X} c(x, y)\mathrm{m}(d(x, y)) \qquad (1)$$

on the set of Borel measures m on $X \times X$ subject to the constraints

$$\mathrm{m} \geq 0,$$

$$(\mathrm{p}_1 - \mathrm{p}_2)\mathrm{m} = \mathrm{r}$$

The mass transfer problem is an extension of the old "déblais et reblais" problem posed by G. Monge in 1781 in his "Memoire sur la theorie des déblalis et des remblais", and admits some economic interpretations:

1. Given an initial s_1 and a required s_2 distribution of some commodity, one has to proceed from the first distribution to the second at the minimum.
2. Provided the cost of transferring a unit of the commodity from a point x to a point y is equal to $c(x, y)$. In the classical case the cost equals the distance between the points, and so the total work of mass transfer has to be minimised.

The original transport problem, proposed by Monge, studies how is the best way to move a pile of rubble (*déblais*) to an excavation (*remblais*), in other words, to move a mass with the least amount of work. The technique used then was by dividing two equal volumes into infinitesimal particles and put these into correspondence with one another such that the sum of the products of the lengths of the paths joining them by the volumes of the particles being transferred is a minimum.

These notions are connected closely to physical concepts such as centre of mass and work. Work is a concept that is related to energy, a concept that is difficult to measure in a concrete way. Instead, we calculate the work done by a force that moves an object from one position to another. In the case of a book lifted above the table, the force is exerted against the downward pull of gravity. Under normal circumstances, work is defined to be the product of the force and the distance through which it acts.

If the force moves the object and this varies in its position, we have to add these partial works and conclude that the total amount of work is the sum. In the continuous case we have that the amount of work: generated by a force from a point a to a point b is:

$$W = \int_a^b F(x)dx$$

Then the work (Energy), can be expressed by:

$$W = \frac{1}{2}v^2$$

If *v* is the square root of the component of velocity, then Work is exactly the norm of such vector in a Hilbert Space. The connection is more evident when we consider that the two more important cases of the uniformly convex cost density considered by [Kantorovich, 1942] is:

$$c(x, y) = \frac{1}{2}|x-y|^2$$

or in other words:

$$Cost = \frac{1}{2}(Distance)^2$$

The other case is

$$c(x, y) = |x-y|$$

or

$$Cost = Distance$$

These L1 and L2 theories are rich in mathematical structure as we can pose Monge's problem as the solution of :

$$I[s] := \frac{1}{2}\int |x - s(x)|^2 d\mathrm{m}(x)$$

in which the equilibrium of masses in a given region is the concept of minimum work.

In modern terms, we are given two nonnegative Radon measures $\mathrm{m}_1, \mathrm{m}_2$ on $\Re^n$, satisfying the overall mass balance condition

$$\mathrm{m}_1(\Re^n) = \mathrm{m}_2(\Re^n)$$

With the condition that the measures are finite.

It appears that the analysis turns to the discourse of measure. Measure is the main concept in the technical matters of Kantorovich's problem. For example a Radon Measure is a Borel Measure that is finite on compact sets, but a Borel Measure is a measure from a sigma-algebra[13] into the real numbers. The importance of a sigma algebra is precisely the concept of measure.

13 A If S is the Borel sigma-algebra on a topological space, then a measure $\mathrm{m} : S \to \Re$ is said to be a Borel measure (or Borel probability measure if normalised). For a Borel measure, all continuous functions are measurable.

The optimisation method.

The optimisation method and the problems related to it were discussed at a meeting of the Mathematics Section of the *Institute of Mathematics and Mechanics of Leningrad State University* in 1939. The paper was originally submitted in that year and, as I indicated, was translated into English in 1960 and published in *Management Science*.

In order to analyse the method I will refer mainly to the mentioned paper and its structure. He divides the paper in nine sections, a conclusion and three appendixes, showing in each section a different problem concerning optimization or in other words finding the optimum or maximum result out of a process (subject to economical conditions). The sections are:

1. The Distribution of the Processing of Items by Machines Giving the Maximum Output under Condition of Completeness.
2. Organization of Production in such a Way as to Guarantee the Maximum Fulfilment of the Plan under Conditions of a Given Product Mix.
3. Optimal Utilization of Machinery
4. Minimization of Scrap
5. Maximum Utilization of a Maximum Raw Material
6. Most Rational Utilization of Fuel
7. Optimum Fulfilment of a Construction Plan with Given Construction Materials
8. Optimum Distribution of Arable Land
9. Best Plan of Freights Shipments.

I will emphasise only on the first point, due to the similarity of the other, however I will discuss some points on the literary analysis and epistemological issues on the whole paper.

The paper begins stating the general problem (which is optimization) and it is clearly about looking for the optimum:

> The immense tasks laid down in the plan for the third Five Year Plan period require that we achieve the highest possible production on the basis of the

> optimum utilization of the existing reserves of industry: materials, labour and equipment [Kantorovich, 1960, 367].

He utilizes the word 'efficiency' linked with the work of an enterprise or industry, nevertheless as seen above, this is precisely the idea of the abstraction of a whole economic system based in the production. He states that there are two ways of improving the efficiency: on one hand, increasing technology such as development of raw materials or new attachments for work, and on the other hand, as he says "far much less used" the improvement of organization of planning and production. About the second way he says:

> Here are included, for instance, such questions as the distribution of work among individuals, machines of the enterprise or among the mechanisms, the correct distribution of orders among enterprises, the correct distribution of different kinds of raw materials fuel and other factors [Kantorovich, 1960, 367].

His language turns to be more of an economist than a mathematician. This might be due to the practicality of the Soviet philosophy at the time.

Bernal, who was one of the "precursors" of the Operational Research groups in England, had been in the USSR in 1931 together with other British scientists; although he might never had any contact with Kantorovich there, he brought from the USSR many ideas in connection with the social function of science. In 1939 Bernal published an influential book on this subject which contains not only the Soviet views, buy also a blend of them with Western ones.

At the tome, the "practical" orientation taken by the academy went beyond intensive planning of all phases of research activities, in order to integrate the operations of the academy into the National Five Years Plan. The government intention was not to force the Academy to abandon "pure science"; but, nevertheless, it wanted the Academy to embark on more practical activities as well - on activities that were measurable by current indices of economic growth. And this social argument could be also a powerful trigger on the approach of Kantorovich's work. According to [Vucinich, 1984, 139]: "In 1932, 60% of the Academy's revenue came from special contracts, a clear indicator of the commitment to "practically" oriented research"

In addition to this, there was an important fact concerning economics.

It is in this environment Kantorovich continues explaining that the methods were developed trying to solve a problem for a laboratory that was in charge of the University of Leningrad.

> In connection with the solution of a problem presented to the institute of mathematics and Mechanics of the Leningrad State University by the Laboratory of Plywood Trust, I discovered that a whole range of problems of the most diverse character relating to the scientific organization of production (questions of the optimum distribution of the work, of machines and mechanisms, the minimization of scrap, the best utilization of raw materials, and local materials, fuel, transportation and so on) lead to a formulation of a simple group of mathematical problems (external problems). These problems are not directly comparable to problems considered in mathematical analysis. It is more correct to say that they are formally similar and even turn out to be formally very simple, but the process of solving them with which one is faced [i.e. mathematical analysis] is practically completely unusable, since it requires the solution of tens of thousands or even millions of systems of equations for completion [Kantorovich, 1960, 368].

In a way he tries to justify science and the discovery of problems in industry and above all mentioning a key word for the higher Soviet circles, as in Western ones, "production". He finds an algorithm to solve the problem without the technological problem of solving a huge matrix or a very big system of equations.

> I have succeeded in finding a comparatively simple general method of solving this group of problems which is applicable to all the problems I have mentioned, and is sufficiently simple and effective for their solution to be made completely achievable under practical conditions [Kantorovich, 1960, 368].

He had the idea of applying this method for the capitalist and socialist system, only being aware of the difference between the goals of profiting in a free market enterprise and the other of fulfil the General State Plan and its component parts, which is to achieve the maximum output and maximum utilization of raw materials.

This discourse is outstanding in the sense of the multidisciplinary approach: a truly mathematical (scientific) note with concrete applications in the economy. Nevertheless, the centre of gravity of these developments was situated in the University environment. It was until years later that the proper incorporation of the scientists in the industrial and other productive sectors was achieved.

The first problem he explains is about the *Distribution of Processing of Items by Machines Giving the Maximum Output under the Condition of Completeness.*

This problem is simplified and, as indicated by the author, it only plays a purely illustrative role, obviously, not to complicate much in the operations but definitely not to do it more pedagogical as this note is clearly aimed to an academic public.

The problem is about the milling work in producing parts of metal items, which can be done by three types of different machines, problem's data is showed in the following table:

Productivity of the Machines for Two Parts[a]

Type of Machine	Number of machines	Output per machine		Total output	
		First Part	Second Part	First Part	Second Part
Milling Machines	3	10	20	30	60
Turret Lathes	3	20	30	60	90
Automatic Turret Lathe	1	30	80	30	80

a. Source: [Kantorovich, 1960, 369].

Thus, during a working day it is possible to turn out on the milling machine, 10 of the first part or 20 of the second part, and so on with the others. The total number of machines is 7 (3+3+1), so the 'Total Output' column expresses the multiplication of the number of machines times the output per machine.

So the remaining problem was to find the distribution to obtain the maximum output; of course, at the same time it was important not only to produce parts I or II, but complete items (consisting of the two parts). Therefore, he had to find the division of work time for each machine in order to get the maximum number of finished items (not parts). And this is where we can find also the complete vision of the method as applicable in the whole economic system, fact that gave a special power in the political and ideological system at that time.

Distribution of the processing of parts among machines[a]

Type of Machine	Number of machines	Simplest Solution		Optimum Solution	
		First Part	Second Part	First Part	Second Part
Milling Machines	3	20	20	26	6
Turret Lathes	3	36	36	60	-
Automatic Turret Lathe	1	21	21	-	80
Number of complete sets		77	77	86	86

a. Source: [Kantorovich, 1960, 369].

It is clear that he has two problems ahead: the completeness and the maximal, however his strategy was first to solve for completeness and not maximal. This is not trivial, because to solve a problem with many variables (correlated among them) one must lead with the correlation between them, i.e. if one alters one of them this could change the conditions of the others and so on until the problem and the possibilities are so great that it turns impossible to solve it.

Then he generalizes to n times m variables and complicates the problem nevertheless, he is able to compute the initial problem with several constrains such as electricity, another utility or labour time from the workers, and then he was also able to maximize not only the output, but also the profit or minimize the cost.

> Further there is possible a variant of the problem in which the production of uncompleted items is permitted but the parts in short supply have to be bought at a higher price, or surplus parts are valued more cheaply, compared to complete items, so that the number of completed items plays an important role in determining the value of the output [Kantorovich, 1960, 373].

Kantorovich remarks that this solutions are applicable to practical life; nevertheless, "depending on the complexity of the case, the process of solution can take from 5 to 6 hours", which after the evaluation cost-profit seemed to pay off. In general, the method solves an economical problem concerning organization and planning production in the abstract turf of mathematical modelling. An interesting note is made by pointing out that further applications should be carried out and that there should be some more investigation concerning the interdisciplinary work with mathematics and production workers.

> Further extensive researches still have to be carried out by the combined efforts of mathematicians and production workers" [Kantorovich, 1960, 387], and continues saying that "In the future it remains to determine the sphere of application of the method, to indicate further problems solvable by it [Kantorovich, 1960, 387].

This tone of discourse aims to build a reality principle in other areas and for other readers and users of the method. In other words, Kantorovich is not trying to give a scientific vision of the problem, but trying to give a reality principle. Some objections to the method were made, but he gives answer to the objections, some of them quite naïve and easily answered by Kantorovich. His answers are more scientific than politic (except for the last one), nevertheless, there were some political and economical objections to his method than mathematical, being the principal the one of mathematics leading with economics.

> We realize that in individual cases it is possible for these objections to be so well founded as to force our withdrawal from a certain filed of application. However, along with these special individual objections, we have been required to counter (in spite of the extremely favourable opinion of the majority) occasional objections of a general character which essentially lead to denial, in principle, of the possibility of using mathematical methods in technical-economic question in the field of organization and planning [Kantorovich, 1960,389].

The first objection is that the theoretical models and in particular mathematical, could not be applicable to technical questions. Nevertheless, he says that the models are very useful to develop experiments, calculations and designing. And that if there is a result led by a theoretical model, which cannot adjust to the condition of the problem, one could perform the minimum changes and adjustments to give a practical optimum solution, out of the theory. This objection was likely from an economist that could see his working area in danger of invasion from the scientists specially the mathematicians.

The second objection was that not all the data could be available, and he answers that the data should be worked out in the enterprise as a normal issue in order to organize it. This idea was developed in the modern planning strategies nowadays, like in Michael Porter's methods and strategies for enterprises. And that if the enterprise lacks of those data, even the minimum planning could not be performed, due to the primitive mismanagement that should be eliminated before trying to optimize its processes.

The third one was that the original data could be wrong and that if that happened all the calculations and results would be wrong. And he says on this, that perhaps in one of the cases this could happen, however, thanks to statistical factors, in the generality of cases this is not what happens.

The fourth objection is that in many cases changing from the ordinarily chosen variant to the optimum only improves the output process in 4% or 5%. He says that the method if applied to all the branches of the economy, not only 1% but a tenth could mean a great result, because it is applied to huge amounts.

In the third and fourth objections we can observe that the model and the way to solve it was so definite, that only minor objections (perhaps very administrative) could only be done.

And finally, the fifth objection is that maybe the results given by the method, could lead to a political or structural movements in the organization political institutions or the society. And he answers that if this is for the good of the national economy, the changes in procedure will certainly be made.

This method, as well as other in applied mathematics allowed translating specific problems and situations into an abstract model, which structure in the pure mathematical world, was the same. This property, well spotted by Kantorovich, was called by him "Universality and flexibility of the model" as its structure permitted various forms of applications, "it can describe very different real situations for extremely different branches of economy and levels of its control" he added when talking about it, (see [Kantorovich, 1989, 20]

In those times, the problem of managing the tool without the computational devices that we have today, was a factor evaluated by Kantorovich: linear models as matrices, (in this case a matrix must be seen as an array of numbers more than as a linear transformation, do to its operational character more than to its functional or analytical character, moreover the discourse was aimed to a non-mathematical audience) can be managed by people with very elemental mathematical training. It is important to note that managing the tool does not imply its understanding. Moreover, the model developed special technical algorithms to solve it.

> The urgency of solving extremal linear problems implied an elaboration of special, very efficient methods worked out both in USSR (method of successive improvements, method of resolving multipliers) and in USA (well-known simplex-method of G. Dantzig), and a detailed theory of these methods. An algorithmic structure of the methods has allowed later to write corresponding computer codes and nowadays modern variants of the methods on modern computers can rapidly resolve problems with hundreds and thousands of restrictions, with tens and hundreds of thousands of variables. [Kantorovich, 1989, 20]

This possibility is given by a system of indices for activities and limiting factors which is found simultaneously with the optimal solution and is in accordance with it. Thus these valuations gave an objective way of calculating accounting prices and other economic indices and a way of analysing their structure.

The success of the method transcended economic systems as it was consistent for many applications, problems and situations of human life. Evidence of its efficiency is the concrete extrapolation to economic problems and Operations Research. These problems began to be investigated in special large research institutes, like the *Central Economic-Mathematical Institute* in Moscow (headed by academician N. Fedorenko) and the *Institute of Economic Science and Industry Organization* in Novosibirsk (headed by academician A. Aganbegjan).

The linear model has proved to be a good means of giving the simplest logical description for problems of planning control and economic analysis and it has contributed to significant advancements in pricing problems. Nevertheless the same Kantorovich spotted some difficulties, firstly with respect to the reception that the theory had had in those times:

> In the most complicated and perspective problems, such as those of national planning, have up till now effective and generally acceptable forms for the realization not been found. The attitude to these methods like to many other innovations went sometimes from scepticism and resistance through enthusiasm and exaggerated hopes to some disappointment and dissatisfaction [Kantorovich, 1989, 21]

In fact, real decisions and the performance of local bodies were not evaluated by the theoretical indices but by actual prices and methods of assessment which are not so simple to replace in a mathematical model. In this sense the applicability of the model into human action was far from achievable.

Secondly, he points out the problem of the correlation of the model with the real world, in which he considers the necessity of experimental science to corroborate results and emphasises on the observations of the objective reality in the construction of data tables. As he says:

> The models emphasize only a few of its aspects and take into account the real economic situation very roughly and approximately, so as a rule it is difficult to estimate the correctness of the descriptions and inferences. It is especially important to test the influence of the difference between the model and reality on the obtained result and to correct the result or the model itself. This part of work is not often observed. The hard thing in a model realization is to receive and often to construct necessary data which in many cases have considerable errors and sometimes are completely absent, since none needed them previously. Difficulties of principle lie in the future prediction data and in the estimation of industry development variants. [Kantorovich, 1989, 21]

Finally, the last difficulty is referred to the computation of optimal solution. In spitof the presence of efficient algorithms and codes practical linear programs were not too simple since they were very large. The difficulties grew significantly when the linear model was modified by any of its generalizations. Nowadays this is not a problem due to the advancement of computational devices and programs to solve such algorithms.

The view from America: the Obituary of the Boston Globe

This obituary situates the image of Kantorovich as one of a single breakthrough in life. It is clear that the image of these methods were not popularised even in the academic core of the United States where the proper figure of Kantorovich or his methods did not reach, possibly for political reasons, the place he deserved. The obituary note begins with the heading saying: "Leonid Kantorovich, 1975 winner of Nobel Prize for economics. Date: Sunday, April 13, 1986" [Page: 79]

And continues with a phrase that might be out of context, due to the fact that what he did could never be considered as econometrics. In fact, he did not develop statistical models applied to economy that include some econometrical models like regressions. In any case, the recognition they refer to might be of other kind, different to the one that was already given by the academy of science in the USSR and the University of Leningrad.

> Mr. Kantorovich was considered the father of Soviet econometrics -- the application of mathematics and statistics to economic problems -- but was slow to receive recognition at home and internationally.

The general tone of the political discourse since the early 1960s was of a comprehensive view on both economic and political regimes. And also science in those times had a position in their discourse above the economic systems, and they pointed out the wide application of the methods to any problem in any economic frame. For example [Gardner, 1990], [Charnes and Cooper, 1962, 255], even writers like [Ward, 1960] who published in *The Journals of Political Economy,* [Kircher, 1961], [Siroyezhin, 1965], or [Johansen, 1966].

> Mr. Kantorovich's first work on econometrics was published in 1939, but was not used immediately because the field had evolved in the United States and was not held in high regard by the Soviets.

In this sense it seems like Kantorovich ideas, if they were good, served as the basis for reconsideration of the political regime of the Soviet Union, to be transformed to a capitalist system like in the United States, which definitely is not the case, as we can see it from his papers, books and lectures.

In fact the American side was not so radical in the scientific circles, as we can see good opinions of high scientists in the field on the work of Soviet scientists, having strong communications with them independently of political regimes or economic systems.

This work might clear out some of the aspects of the development of applied mathematics in what we know today as Operational Research and the persons who are the main characters and precursors of these theories.

The nature of linear programming.

Dantzig attributes in his [1963] the name "linear programming" to Koopmans (as an alternative form of his "programming in a linear structure"), and as stated by [Grattan-Guinness, 1994, 62], in a later version he also credits R. Dorfman with devising the name "mathematical programming". The nomination might be an important problem, because it also says something about the methodological nature of linear programming. Its direct applicability makes it a useful method more than a heavy theory of functional analysis that the user has to learn. In fact, nowadays, many learners do not know the details behind the method, but the method itself and how to apply it, however, at this moment I am focusing on the discussion of some of the technical developments in the discipline and a possible overlapping with the computing methods.

In those times the new computational devices and techniques made more accessible the solutions proposed only theoretically by linear programming models.

A leading figure in mathematics and in computing, J. von Neumann, made some contributions in this field, linking linear programming methods with the theory of games. By the time of von Neumann, operational research techniques were being vigorously adopted by military and industrial institutions to solve practical problems concerning the decision making of projects. Nowadays, operational research contemplates under its name linear programming techniques, statistics, experiments, modelling, stochastic processes, computational techniques and other methods mainly developed in this time.

Although Kantorovich was also concerned about the computational methods to achieve a concrete solution, he was aware that the essence of the linear programming problem was not a computational one, in fact, we can say that epistemologically, the difference between computing and linear programming is radical; if one the one hand methodologically, linear programming and computing overlap, on the other, in the epistemic sense, their roots and development are different.

Statistics, linear programming, and in general all the areas of Operational Research including the areas of management explored by Kantorovich are, first of all empirical methods, for empiricist, cognition begins in the process of experience, of course experience with respect to a subject; therefore the first epistemological reflection revolves around the place of knowledge in reference to the cognitive subject. In computing the cognitive subject has a different kind of cognition.

In linear programming methods, the cognitive subject is the one controlling the variables to optimise; the cognitive subject is also controlling the constraints of the problem. In computing, the cognitive subject develops a solution following an algorithm. In other words, cognition in computing methods is achieved by the syntax of the code and its interpretation in the sense of several instructions, meanwhile in linear programming, cognition is achieved in the processes of controlling, which are directly related to the rhetoric behind its discourse.

Kantorovich's discourse on linear programming revolves around the concept of measure control, (as the optimum is a measure as well); this happens also in other cases of Operational Research methods. Moreover measurement and measure theory are among the theoretical basis of statistics as well. Thus, measurement in the experiential and epistemic sense becomes a main issue in the discussion.

Measure in this sense has to be done by someone that might or might not be the same originator of the discourse, as a counter example we have the case of Blackett, in which the generator of the measurements in the experiential level was not Blackett.

The narrative identity versus the cognitive subject in business development.

The first object of the discussion is to define what meaningful knowledge in applied science is, and the natural way to do it is following the epistemic way of the discipline revising the literature in it. As seen above, Kantorovich, and other pioneers of the area have developed not only the methods in a mathematical way, but also a speech in which the explanations and developments are given. Furthermore, this discourse is given in different context (as in different political and economic regimes) acquiring a special status of scientific narrative, or what Lyotard calls the "Great Narrative".

The cognitive subject in the case of applied sciences is clearly not the data owner, if it was, then a subject like Operational Research (including linear programming) could be reduced to a mere technique to manage data in which meaningful knowledge does not exist.

In this case it is the analyst who constitutes the first cognitive subject, i.e. the cognitive subject is not the general public, the government, the military or in general the data owner as might be in other cases, but the mathematician (analyst) himself as he is doing an internal hermeneutic as he reads the outputs of the process; and it is in the mathematical process in which the generation of knowledge takes place, because he is the one that is generating the general narrative of the subject and also making the first interpretations, explanations of the models, launching a reality principle for the reader (second cognitive subject) of the discourse.

In the case of Kantorovich concerning his contributions to the optimal allocation of resources, one can see that the epistemic characters of narrative identity and cognitive subject overlap.

In Kantorovich;s case cognition is two-folded due to the nature of narration. Narration has to poles as in any communication process: the emitter and the receiver. An important analysis is to identify who are they and what are their respective positions in the process[14]. The main difference to identify these characters can be found in the process of representation, as seen in the case of Fisher (Chapters 4,5, and 6).

Applying this concept to Kantorovich's case, concerning his activities in applied mathematics, the <narrative identity> is characterised by its constant evolution in many planes, the first plane is the proposition of a model in which the variables are chosen, the second a method to solve the mathematical problem, another plane would be the general algorithm as a method, (this step is important in the dynamic hermeneutic sense, because it generates (transforms) a methodology out of the epistemology), one of the last planes is the generation of a second cognitive subject and a new narrative identity as such. A huge epistemic[15] problem in Kantorovich's work[16] can be delimited.

In Kantorovich's work, one can identify the beginning of the movement towards a possible world (between real world and its mathematical abstraction), produced in a relationship between this narrative identity and the cognitive subject, but only in the pursue of, not the principles of nature or origins of a production economic theory, but of a writing activity in the sense of the construction of "expecting horizons" as in the case of Blackett.

In the case of Kantorovich the determination of a reality principle taken from his discourse and its epistemic procedure, is linked not with a representation of reality, but with a production of a "common sense" taken from the same narrative, from which "anyone" can discern a principle of the real. This is not an application, but a reflection towards an applicatory action.

14 If we want to be precise with the use of language we can say that applied mathematics has processes and procedures. Discussion on the difference on these terms can be found in Saussure's *Course General de linguistique* (for an English version see [Saussure, 1959, 176]).

15 Here epistemic refers to the creation of concepts and creation of planes, See [Deleuze/Guatarri, 1991].

16 In general this can be applied to other cases of applied mathematics.

As said by [Gardner, 1990, 646]:

> Kantorovich said, "A major achievement of the mathematical economic direction was the elaboration of a series of problems of planned pricing, as was the sustentation of the thesis of the inseparability of the plan and prices" (Kantorovich, M. Albegov, and V. Bezrukov 1987).

The discourse surrounding linear programming methods gives the generation of knowledge in the field a unique characteristic, thus allowing the analysis and delimiting well the reaches of the discipline as a cognitive rhetoric.

Chapter 8
Latin America: A Case Study for Strategic Planning.

Introduction.

In this chapter I will analyze some cases in Applied Mathematics in Latin America and analyze the similitude between the European cases analysed in this work, (Blackett-Fisher-Kantorovich) in the context of modern business environment.

When businesses fall into complicated environments, for example, when they can be non-regulated with respect to competitors or at times very constrained due to special law or market circumstances; or moreover when politics and cultural awareness are main factors, then Latin American cases can be taken as leading cases for business development.

Good lessons can be learnt out of emerging markets as treated independently, however in the scope of this work I will analyse a generality of circumstances linked to the epistemological frame of applied mathematics.

Along this book I have outlined the general epistemology of applied mathematics to business in possibly the most important cases in the history of the discipline. The general cognitive space is framed by a constructivist philosophy, however, as we saw in chapters one to three, the treatment of images and the theory of language are of the utmost importance along with the methodology used.

In this chapter I will only mention the methods, as they have become of general use, studied in many business and applied mathematics specialisations and are vastly documented in literature (books and papers). But I will analyse their epistemological features instead.

Before continuing, one important distinction has to be made between business development and business evaluation; although at a cognitive level they might be similar, we have to outline properly that there is a different epistemology between the evaluation of, for instance, capital investment projects and the generation of strategies based on scientific (or mathematical) methods. Strategies in the epistemic level are constructed based on the communication processes within the same organisation; in other words, semiotic and hermeneutical issues lie beneath the analysis and development in and out of business and this sort of enquiries should be performed as in the cases of Fisher and Kantorovich to understand the proper dynamics of business development.

Mathematical models such as linear programming are also applied to the organisational structure itself to optimise its performance (as a team), however their proper implementation and moreover their proper application to businesses relies on language issues.

This chapter is the most applied one and it is a result of the conciliation of many theories and authors based on a personal experience as a business developer. It is not concentrated in one stream of thought or in any special theory to explore problems related with the generation of a proper evaluation of business or their development, but it is rather epistemologically based.

Indeed, there is a historiograhical proposal and an epistemological one, which are the main conductive thread in the structure of the analysis. In fact, not only Latin American or international businesses are special cases, but nowadays, almost any business project has a unique circumstance and a unique epistemology.

Obviously to construct a sui generis epistemology for a specific case of study has its disadvantages, especially when one tries to present the work to a rather orthodox audience. In this sense, this area is not very visited by many academics that prefer to take a comfortable position and adscript their work to a formal and already recognised model. In this case I am more interested in an independent epistemic construction in which "epistemology" is taken in the Deleuzean way meaning to draw planes and create concepts.

This work is written to show that the writing activity is a powerful tool in the making of management science and business development, without diminishing the power of mathematical modelling and other theories.

The relationship between thought and writing and between perception and though is analysed in this part as one of the most important steps in business planning. That is why we stress our argument in what we consider strong epistemic steps, like representation, in which we try to articulate its inherent antagonisms in the materiality of the real world and its applications. In other words, representation creates a space for which the conception of images plays a main role. In this chapter we try to explain its importance and ways of working in the making of a business development.

It is important to state that the linguistic, philosophical and mathematical resources are support for a managerial and business-wise one, in which methods for modeling and evaluating are launched. Sociolinguistic analysis and linguistic analysis are not the main goal of this work, but only the production value based in the exploration of the epistemic limits of this approaches.

An epistemological frame, given the conditions of contemporary though constitutes a necessary condition to establish these limits and borders, which have been traditionally taken for granted.

In applied management, the writing activity creates an epistemology or a cognitive situation, in which the text is the main unit; this text leads to a kind of philosophy which models and explains thought in a convenient way. As in business planning thoughts are constitute a prior part in the planning-action duet. In this case, a philosophical frame is taken from the ideas of the constructivist theory, for which knowledge is a construction of the human mind and in which the writing activity, as token of communication, plays a main part.

In this case the constructivist theories applied in this work the ones of Maturana and Varela play an important part, Maturana's work is based in a concept called "the tree of knowledge" (see [Maturana 1984]) in which a problem of language is the base of it, and a branch of it the relationship between knowledge science and a writing activity as and imaginative product (see [Calvino 1993]), other important branches of this tree are the philosophical status of the problems, the historiographic treatment and the invisible reception of the work (comments and reviews). In this sense a comparable frame can be found in [Deleuze 1976] and his concept of Rhizome.

The constructivist theory launches a generic relationship between the biological dimension and the cognitive development, in this way, the launch of traditional cognitive taxonomies is related closely to a physical formulation of writing ideas, in which an open and particular epistemology is proposed.

This work does not try to analyse the evolution of this tree of knowledge in a philosophical way (XIX century) nor the concept of tree of knowledge per se, but as Maturana and Varela expose, it is the intention to explore each topic and question in terms of a tree of knowledge putting aside any universalistic ambition as tried by other academics, who like to base their historiograhical work in one theory.

In this sense, this chapter gathers the relation between language, knowledge and applications to the "real world" of specific models of applied mathematics, placing each generic question in its conceptual totality.

This work also faces in the first place, the essential difference between science and technology concerning business planning and development, that can be depicted, as the reader may suppose by now, as a cognitive problem, in which the applications of knowledge are not figurative, but linguistic (without disregarding a material dimension), in which questions of cognition turn to be more operative that conceptual.

In this sense, a cognitive development as seen in the major three cases analysed in this work, are produced in a local manner but generalised in modern business frameworks, and the perception and experience create a construction of knowledge in which a representation (in the form a map or as a cartographic activity) is the first epistemic step.

Our historiographic position contemplates and extensive and complex elaboration that is a necessary condition to lead a platform for an epistemic debate. In this sense our historiograhical proposal considers a sociological analysis and contemplates the conditions of physical and imaginary circulation of the texts. The historiograhical position is contained in the philosophical one, in which we analyse the system of beliefs (business framework) in which this circulation of texts is done.

Summarising, in this chapter I will 1) outline the formation of the notion of a cognitive subject (a business developer), such as the one proposed by J. Piaget, inseparable from the creation of an idea of knowledge, 2) analyse any form of knowledge as a construction of local characteristics of "local knowledge" (subject and environment), 3) explore the construction of a space in a world in which the mentioned subject and the notion of possible knowledge constitutes not only the fundamental movement of a reality principle, but also a continuum in that relationship as a principle of change. This is a basic characteristic for the generation of business cognition (business planning as well) and 4) determine the formalisation and construction of a space (necessary for any notion of subject and idea of knowledge) which demands a theory of language and writing.

Research models on Project Evaluation.

As Mary E. Beadle (see www.mckinsey.com) says in her article "The Influence of U.S. Media Use and Demographic Factors on Argentine Men and Women About Perceptions of U.S. Lifestyle" "All international business activity involves communication" (Martin and Chaney, 1992, p. 268), and an important factor of achieving results in modern companies is the fact that we can make communicative bridges between the mathematical and/or analytical models and the day-to-day language and Latin America is not an exception. In the case of Latin America another important issue is the fact that we have a strong tradition of models copied from the U.S. and Europe, so translations and language issues are adapted in a different way to certain situations.

In her report on how American media influences Argentinean businessmen Beadle tells us that "Communication across cultures is difficult because it includes more than language". And that "U. S. firms have had between 45–85 percent of their expatriate U. S. citizens return early from foreign assignments because of their inability to adopt to a new culture (Martin & Chaney, 1992)".

Indeed as in the case of Kantorovich, perceptions (or the reception of a recommendation, document or report), are the first epistemic step in the communication process. And this might be the reasons why this 45–85% of U.S. citizens return early from foreign assignments.

Understanding how these theories and models are received in other cultural environments other than where they were originally developed has different steps.

For example in the case of PEMEX in Mexico the constant aid of McKinsey, Curtis Mallet and other international firms applying foreign models has marked a clear influence in how Latin American big corporations develop business.

For example the reception of theories or the reception of the same scientists in Latin America like the case of Birkhoff or Einstein in the middle of the 20th century was marked by a strong political influence. In fact, it seems that the backup of an international name is as important as the analysis performed.

Contrary of what Beadle says, intercultural communication in the international business environment has a different flavour in terms of scientific analysis. In the case of PEMEX the usual models are set in an European tradition of science and a North American analytical practice. Indeed I agree that studies quote by her indicate differences in perception as to how U.S. citizens and foreigners see U.S. however in terms of methodologies, the emulation is clear.

According to Beadle Ruch and Crawford (1991) report that in general, Latin American business cultures prefer face-to-face communication, in which previous analysis had not been done in advance thoroughly. And this follows exactly the case of Garcia in Peru or Sandoval in Mexico in the decade of 1930.

Vicente F. Assis, Bernard Minkow, and André Olinto in their report entitled "National oil companies: The right way to go abroad", published on November 2005 (see it at www.mckinsey.com), state that "Often, strategies were poorly crafted, so the companies were slow off the mark or lacked the necessary resources (including technology and talent) to accomplish their goals". What they tend to do including the case of Mexico is use joint ventures with international petroleum corporations to fill gaps in skills or technology or as said by Assis "by improving the process for recovering oil from existing wells"

Ideas that come out of this approach to business can be useful in helping us to understand the modern world: its institution, its problems and its trends. Mathematical theories have been widely used in finance, production, transport or war but these theories have been lost at the expense of the mere technical tools in order to mechanize thinking processes to "solve" problems in human life.

This loss may have been due to the common application of these tools or in order to make them reachable for many. Another possible reason is that universities nowadays have to deal with corporations and have to respond to the productive sectors of society that demand certain kind of people for their organisations: the financial and strategic considerations have replaced the purely pedagogical. This means some universities nowadays manufacture technicians, to deal with practical problems in real life. Philosophical discussions are not part of the syllabi of many institutions anymore, not even for doctorate purposes.

However, due to the speed at which many companies have to move, managers and directors have to take decisions quickly with limited information, and it is understood that many of them base their decisions on the analysis produced by a team of experts like the operational research sections in military organisations (chapter 3); this currently fashionable situation of institutions and corporations being reliant on their team of advisors has lead to the flourishing of companies of technicians calling themselves consultants selling their services for fees.

In politics, and generally in industry, decisions and discussions revolve around methodologies, nevertheless epistemological grounds are never discussed, even when discussing an up-to-date problem such as knowledge transfer or education policies which is clearly epistemological, the methodology preserves most. What this means is that some companies are spending their time looking at the logistics of a programme without questioning its rationale.

Looking at these issues one has to have someone able to question the fundaments of what we are doing, but universities hardly produce these people: it's quicker and easier to produce thought-free technicians who are trained to apply a formula. For example, traditionally analysts research on the performance of companies from an accounting perspective. Based on balance sheets and accounting results (quantitative inputs), classical financial ratios and forecasts are performed and the evaluation of the company is done. In fact this analysis, considered being the best we have, most of the time has errors and does not allow for qualitative variables such as human or social behaviour.

I would like to make a parenthesis here to comment on what I call traditional methods. In 2004, one of the most important universities of the United Kingdom, the majority of the PhD theses in Applied Mathematics (that is statistics, finance, operational research and mathematical physics) work either in computational solutions for differential equations or deal with models that in the best case were developed in the 1950s like the Monte Carlo method and its variations such as Markov Chains Monte Carlo, better known among technicians as MCMC.

While performing historical research in statistics I realised that models proposed in the 1930s are currently used and are considered to be the state-of-the-art tool in many fields nowadays. For example, in the design of experiments, concepts like the correlation coefficient and the algorithm of linear discriminant are widely used in credit scoring, another example is the so-called data mining.

The most relevant issue is that the underlying methods performed in applied mathematics, come from a philosophical discussion made by the precursors of these theories like R.A Fisher, L.V. Kantorovich, P.M.S. Blackett, or Dantzig only to name some.

In the development of fields like statistics, medical trials, and operational research, one can see that main characters in the field like those named above, were concerned on a wider view and not only a minimalist view of the numerical results.

In fact, following the epistemic processes of these breakthroughs in applied mathematics, modernly used by consultants to optimise logistics, and other processes in companies we can realise that the analysis began with a wider approach, often not numerically translatable; this is clearly the case of P.M.S. Blackett and the emergence of Operational Research.

This is a consequence of an exploration on the epistemological approach of some analytical processes concerned with mathematical tools in order to evaluate processes and overall performances in modern companies and institutions.

The extrapolation of these meaningful cases in history to modern situations can be successful in practical cases as it contemplates a multidisciplinary view of a traditional task taken in modern enterprises.

Sometimes the analysts hide behind their technical knowledge and shield themselves in the argument that if one does not understand the models or algorithm, then one has no right to make an opinion about the problem under analysis. In many cases, in my experience when I asked them to explain my the model, they did not know how it really worked, confirming my suspicions that they are merely technicians like the driver that gets in his car, but does not know how does it work. Indeed sometimes, they can drive very well, but when there is a problems or when one asks them about the possibilities or limits of the car, they simply do not know. That is why this kind of analysis is of the utmost importance if one wants to analyse the essence of business development.

Mexico's Oil Market: PEMEX

Parts quoted in the following text are taken from the official PEMEX source and can be consulted in its website.

Background: Since 1986, Mexico has privatized or eliminated more than 1000 state-owned companies; today, approximately 252 companies remain in state hands. Mexico's privatization efforts are currently concentrated in seven principal sectors: roads, railroads, airports, seaports, telecommunications, electrical power, and natural gas. Limited foreign investment in Mexico's energy sector is possible.

Mexico's Five-Year Energy Plan: 1995-2000: On February 12, 1996, Energy Secretary Jesús Reyes-Heroles unveiled a five year program to develop and restructure its energy sector. The plan, known as the Program for the Development and Restructuring of the Energy Sector 1995-2000, outlines steps Mexico must take to improve the quality and distribution of its products, spur technological development and expand production while creating a more balanced approach toward the regional development of resources. The long-term goal of the plan is to increase the private sector's investment in the energy sector so as to improve the overall strength of the Mexican economy. According to the program, Mexico will need a minimum investment of approximately $33 billion over the next five years to meet its goals, with approximately half that amount invested by the private sector.

To encourage private sector investment, two key subsectors of the energy sector will be developed and privatized: natural gas and electric power production. Despite previous plans, another subsector, secondary petrochemicals, will not be privatized and will continue to be under PEMEX's control. However, minority shares in secondary petrochemical plants may be sold.

Natural Gas: It is estimated that Mexico will need $3 billion over the next five years to upgrade and expand its natural gas sector. On November 8, 1995, Mexico published regulations opening the natural gas sector to private (Mexican and foreign) investment for the first time since 1938. Foreign investment is permitted up to 100 percent in the transportation, storage and distribution of natural gas and up to 49 percent in the construction of pipelines and drilling wells for oil and gas, subject to increase with the authorization of the Mexican National Foreign Investment Commission.

Electricity Generation: In December 1992, Mexico changed its basic law on electricity to permit the participation of the private sector in power generation not for public service. Implementing regulations were issued in 1993 and modified in 1994. Private investors may construct, own, and operate generation facilities for the purposes of self-supply, co-generation, independent power generation, and small production (up to 30 Megawatts of capacity).

By the year 2000, Mexico's electricity needs were expected to increase 12,000 megawatts to a total of 43,000 megawatts. To meet growing demand, the CFE is expected to allocate power generation contracts worth $6 billion over the next five years, for a combined public/private investment goal of $10 billion. Furthering this goal, Mexico's Five Year Energy Plan announced the Government of Mexico's intention to auction concessions for private sector construction of 21 new power plants over the next four years.

Secondary Petrochemicals: The Mexican government announced in October 1996 its decision to abandon the long planned privatization of its secondary petrochemical industry, offering instead a large stake to the private sector. Rather than fully privatizing the 61 secondary petrochemical plants, PEMEX will retain at least a 51 percent stake in each plant. The sale of the first unit put out to bid last April, Cosoleacaque, was canceled. Mexico will, however, seek to encourage private investment in new projects through liberalization measures.

San Fernando Pipeline. This is a clear example of the type of business developed and analyzed in the business planning division of PEMEX. This was a $226 million project; a 114km natural gas pipeline now operating in the State of Tamaulipas. This project was sponsored by Pemex- Gas y Petroquímica Básica and El Paso Energy Company, through a unique Mexican joint venture company.

The pipeline was to provide 1Bcfd of gas to power 15 proposed new power plants on the Gulf of Mexico in Altamira, Tuxpan and Tamazunchale. Once in operation, the pipeline will increase Mexico's overall gas transportation capacity by almost 20%. (The transaction was named the 2003 “Latin America Oil and Gas Deal of the Year” by Project Finance magazine.)

Legal, Regulatory and Other Issues. As a result of these and other activities, One, as a consultant, has to have a good understanding of the legal, regulatory and political issues that affect both power and gas pipeline transactions in Mexico. They are familiar with the principal governmental entities, including CFE, PEMEX, the Comisión Reguladora de Energía (“CRE”) and Hacienda (the Finance Ministry).

They have a history with each of these entities that extends back several years and history on important transaction-structuring issues, such as a new proposed pricing and regulatory structure for sales of natural gas and the regulatory requirements for the construction and operation of new natural gas pipelines. These issues should be contemplated as a sociopolitical variable.

PEMEX's projects are of the utmost complexity, which involves financial problems, business arrangements and ad-hoc schemes, transportation problems and other operating arrangements

Finally, the political and business arientation in the case of PEMEX is in line with the following builletin of the 15 May 2006.

> La fusión de las empresas petroquímicas de pemex permitirá aumentar la eficiencia y reducir costos A partir del primer día de mayo, seis empresas filiales de Pemex Petroquímica (PPQ), que funcionaban como sociedades anónimas, operan nuevamente como complejos petroquímicos, y una más se convierte en unidad petroquímica. En todos los casos, sus activos quedaron integrados a PPQ.
>
> Con la aprobación presidencial de este proceso, Pemex Petroquímica podrá integrar la totalidad de sus proyectos y recursos en un modelo de operación que permita superar las debilidades y aprovechar las fortalezas del esquema anterior, así como las de Petróleos Mexicanos en su conjunto, a fin de tener una coordinación mucho más estrecha y efectiva entre los centros de producción y las oficinas centrales, optimizando las funciones de comercialización, operación de las plantas, planeación y administración y finanzas. Por lo que respecta a la actividad comercial, Pemex Petroquímica buscará que ésta sea conducida en forma centralizada y ordenada, incrementando su competitividad en los mercados, a través de la aplicación de tecnologías electrónicas de mercadeo (e-commerce) y mediante contratos de compraventa unificados a largo plazo, con condiciones más favorables.
>
> De este modo, será posible establecer una organización orientada al mercado y al cliente en particular; crear mecanismos de coordinación entre las áreas comercial y de producción; buscar un régimen de operación comercial flexible que permita un desempeño ágil en un mercado dinámico y altamente competitivo, así como rediseñar procesos de negocio. En materia de producción se busca generar sinergias por la operación conjunta de todos los centros de trabajo (instalaciones, inventarios, compras, recursos técnicos especializados), optimizar la cadena de suministro entre proveedores-procesos-clientes, mejorar el cumplimiento de compromisos comerciales, y disminuir costos de operación.
>
> Con la fusión de dichas empresas, la planeación estratégica se orientará hacia un mismo objetivo, a la vez que se desarrollarán proyectos encaminados al mercado para aprovechar las cadenas productivas existentes, creando estrategias integrales para mejorar los procesos productivos y esquemas de coinversión.
>
> Asimismo, se logrará eliminar las desventajas en el manejo financiero del capital de trabajo de la empresa; construir un nuevo modelo de operación basado en el uso eficiente de la tecnología de información; estandarizar y mejorar los procesos administrativos, y sobre todo, reducir gastos de administración. De esta

> manera, ya operan como complejos petroquímicos: Cangrejera, Cosoleacaque, Morelos, Escolín, Pajaritos y Tula, y como unidad petroquímica, Camargo.
>
> Estas estructuras están enfocadas a la actividad de producción, en tanto los demás órganos administrativos operarán con dependencia lineal y funcional de Pemex Petroquímica, y como soporte a los procesos de comercialización y producción.

So we can see that the business planning is in line with the concept of information exposed by Fisher, for example, the administrative processes are achieved after the analysis of business structures and not the other way around.

Here one of the most important issues is the variable control, convergence and other concepts from the so-called hard sciences.

Variable control.

The vast majority of the problems in operational research or management science are already minimised into some parameters that must be optimised, nevertheless, before that evaluation on control of variables and, above all, the evaluation of non-quantitative variables is always dismissed.

Sociological, psychological, and even philosophical components of the problems, situations or projects are not taken into account because they are not easily quantifiable.

Now, the real problem, which is not treated in any classical book is the variable control, because it is more a philosophical issue that develops into a mathematical model that begins basically with the problem of how to observe, and one cannot teach someone else how to observe, as this is totally subjective. The pedagogic example begins when the variables x and y are already given, and the only thing we teach in the classroom (form secondary school to postgraduate studies) is the technique of solving given specific problems. In other words, if one opens a book of applied mathematics, for example [Ravindran, 1988] on Operational Research one can find that the problems are already minimised to a standard form, the variables are given and the algorithm is given.

In real business the simplest and quicker analysis are generally the most successful ones. Managing and choosing variables comes only by processes used in studies like cosmic radiation or in the epistemology of experimental design, in which variables are always to be determined.

In PEMEX, the latin American case that we will analyse, the amount of variables were sometimes out of control (no information or volatility) and only via iteration of processes one could keep controlling and evaluating them. Business analysis in general should begin with the exploration of variables, in other words choosing the units of measure.

Convergence and Quick Convergence.

One of the most important concepts in mathematics is convergence. Convergence is treated as a main topic in Calculus, Functional analysis [Kantorovich, 1953], and in mathematical methods for Physics [Courant-Hilbert, 1953]. Convergence can be seen in two parts: convergence to a number or convergence to a function. In applied mathematics usually one has to deal with huge amount of data and the aim is the reduction and classification of them. A very useful tool for this purpose is the so-called correlation coefficient, which in his first times had the spirit of variable control.

The other is the unification of concepts. As in [Kantorovich, 1953], series of numbers are put in polynomials, which can be understood as vectors or matrices and then converted into functions. From these functions derivatives and other operators can be obtained and interpreted in an optimal way.

In this sense, a series of data (not necessarily information) is converted into a formal element of a space or of a structure in which the notion of information takes place.

In a way it is the understanding and further interpretations of the simplest ideas the action by which meaningful breakthroughs in business science are achieved. Quick convergence to functions or numbers are of the utmost importance even in the historiograhical cases analysed in this work. Due to the dynamics of the markets

PEMEX executives had to rely on quick convergence in order to take optimal decisions "on time". To do that one has to rely thoroughly on hermeneutical techniques to translate quick and accurately mathematical models.

For example, Fisher's books are far from scientific papers, they are closer to technical business works, and sometimes are nearer to real and everyday problems. This innovative point of view had not been written for the experimental workers hitherto: he converted the experimental worker into a statistician, as Statistical Science has and ends: "the reduction of data".

> In order to arrive at a distinct formulation of statistical problems, it is necessary to define the task which the statistician sets himself: briefly, and at its most concrete form, the object of statistical methods is the reduction of data.[Fisher, 1972, 278]

It does not say reduction to the maximum which one would think could be zero data (or one), moreover, the process of reduction of data can be achieved in many forms and it is totally at the discretion of the researcher (hermeneutical), this could lead to discrepancies in the interpretation of situations (situations quantified by the data) therefore the necessity for a standard method for minimising data, or in other words statistical method emerged. In Fisher there are three forms to do it, which he calls the three main problems of statistics:

> Problems of specification. These arise in the choice of the mathematical form of the population.
>
> Problems of estimation. These involve the choice of methods of calculating from a sample statistical derivates, or as we shall call them statistics, which are designed to estimate the values of the parameters of the hypothetical population.
>
> Problems of distribution. These include discussions of the distribution of statistics derived from samples, or in general any functions of quantities whose distribution is known [Fisher 1972, 280]

The way of dealing with these three problems is mirrored in the epistemology generated in this step. Actually in modern business environments I have used several different techniques depending on the information and the urgency for results.

Models from Physics and Mathematics.

Historically, we often find that the way to extrapolate mathematical models to any area is achieved first via physics. For example an expression that can be interpreted in mathematics as a linear combination can be interpreted in physics as the momentum or centre of mass. The same expression can be also interpreted in other contexts as the mean of certain sample in statistics that indicates the expected value for some distribution.

In many areas of economic analysis a crucial factor for the decision-making is the position of the company with respect to the total market share. Competition and in general all market projections like production and market prices are contemplated in this analysis. One of the main economic concepts to do so is the elasticity of the demand. In oil markets, this evaluation was commonly used and its modeling was multivariational.

It is important to note that for the purpose of this work demand is referred to a multivariate function, that for simplicity (without loss of generality) is represented by a curve or a line in the two-dimensional space.

The formal definition of the price elasticity of demand is:

$$\mathrm{h} = \frac{\mathrm{D}\%Q}{\mathrm{D}\%P}$$

And the cross elasticity is defined by:

$$\mathrm{h}_C = \frac{\mathrm{D}\%G_1}{\mathrm{D}\%G_2}$$

In both cases the measurement is bounded form -1 to 1. Indicating the relationship, on the one hand, between price and quantity and on the other one good with another.

In general this is represented geometrically as the slope of line in a two-dimensional Euclidean space. However, many other considerations should be done in order to understand the principles behind the measurements.

Other example is the rate of comparison in the consumption level under a policy A compared with the consumption under policy B, given by:

$$l = \frac{u(C_B) - u(C_A)}{u(C_A)}$$

where

C_A is the consumption level under policy A,

C_B is the consumption level under policy B,

$u(C_A)$ is the consumer welfare under policy A,

$u(C_B)$ is the consumer welfare under policy B

This expression was derived from the genuine comparison in the consumer welfare as an equation that measures exactly the number of units l that exceeds on policy on some other one.

$$u[(1 + l)C_A] = u(C_B)$$ [1]

In order to know the value of the l one must assume linearity of the function u, which gives us the information and behaviour of a normed vector space. This means that properties like continuity and triangle metrics are valid in the space of economic action. Moreover, this means that the definitions are local in a topological sense and can be generalised only under the assumption of the same dimension.

The locality of the concept of percent change.

Concepts in economics do not float in abstract spaces, they are grounded in some space in which they take form and acquire meaning.

1 This expression was given by Robert Lucas in his presidential address delivered at the 115th meeting of the American Economic Association held on the 4th of January 2003 in Washington D.C. A reprint of this paper can be found in The American Economic Review, Vol.93 Number 1 of March 2003, pp. 1-14.

Following [Deleuze, 1984], concepts are fractal parts that can only live in the context of a plane. This plane of immanence is, in the case of economics, allows concepts to span infinite ideas. It would be wrong to consider that these concepts in economics change. In this case the space in which they exist is the one that changes.

What kind of information are the economic parameters.

Statistics, and other methods of information management play an important role in the interpretation of economic information. But in a contradictory sense, the effectiveness of the method used to derive the information is measured more in the technological part, i.e. the explanation of how the final outcome of a process was achieved.

For example the methods used in neural networks, like the "back regression perceptron" is at the end another dynamical system which consists of linear transformation alternated with non-linear transformations, in order to simulate the way in which the human brain works and to give an effective result of some information given. Nevertheless, the explanations are not as clear as in other methods like the so-called artificial intelligence or the classical statistical models.

Knowledge in economics.

Concerning the cases in which some parameters are measured, knowledge is still given in the form of a great narrative; in fact this narrative is constructed in such a personal way, that there are many interpretations made on the same information.

Organizations in business.

Many parts of the organisation have to deal with analytical areas, business and customer service, engineering and technical work, logistics including materials management, delivery and financial predictions on the life of assets and capital investments, which are in the accountancy area.

In all these different parts of the enterprises, concepts like knowledge, reliance, equality, expertise, mastery and functions, are key to achieve an efficient management on knowledge transfer.

Nevertheless, important questions not often contemplated by management science are the epistemological and technical approach in the process.

Some issues analysed in this chapter are the cultural differences presented in Latin America, and although we are aware that some parameters of measurement are quantifiable and already treated by others researchers, for example the expertise overlap rate, considered in [Reagens, 2003] as:

$$eo_{ij} = \sum_k a_{ik} a_{jk} N_i$$

could have a great impact in the sociological and epistemological turfs, in fact, the above equation can be seen as a linear combination that could be given in a more effective way in the frame of artificial intelligence, considering a binary tree and a machine learning process to evaluate the efficiency of the system.

What is Cognition in business?

In order to talk about knowledge ion business we must first talk about what is knowledge in this frame, i.e. what does it mean to have knowledge and what is the process in which it is achieved, learnt, understood or given.

Knowledge in this case is going to be divided in: i) in the sense of the act of cognition and ii) in the sense of meaningful knowledge in the pragmatic way.

In the first case, the act of cognition in a computer system is completely different to the act of cognition in the human resources organisation or in an internal control of some financial department.

For a technical department cognition could be resumed as symbols and its manipulation, meanwhile in a non-technical department could be taken as a narrative in which a linguistic structure plays an important role in it.

In one case, knowledge aims to be communicative instead of informative; knowledge is encoded in a way to be easily manipulated, as the final aim of cognition is that: the effective manipulation of symbols. For a computer system as well as any other analytical and technical area, for example operational research or economic analysis, knowledge should be given in symbols and the operators should be able to manoeuvre with them in order to get a concise result.

In the other case, knowledge is pragmatic, in the sense that the main goal of it is to be understood. The forms in which it is given are situational and in many cases learnt in a narrative or experiential way.

Experiments or experiences in the day-to-day work should be taken as a pre-epistemic step, in which knowledge is not even formed. Whilst in the technical sections the first epistemic step comes with the representation of information in the system of codes in which a semantic analysis is limited to the programmer's discretion. Indeed, for the technical areas the syntactic part plays the main roll in the knowledge transfer.

In a way, the system of codes comes always with algorithms which in general have no need to be explained or specified to be acknowledged or manipulated. Cognition is implicit in the syntax rules of the language game generated by the technical subjects like mathematics or mathematical logic.

For the non-technical parts there is a whole logistic process which main form is given in a textual way.

Textual and logistic analysis should be performed in order to reach the maximum effectiveness in knowledge transfer.

Logistics

The necessity of manuals like internal controls or books of procedures is effective to one point and for specific results. For example the generation of internal control manuals for the corporate financial departments is generally based in previous experience and its goal is to be applied in a semantic way. Interpretation of this kind of manuals is crucial in their application.

Textual analysis.

Textual units must be identified to generate the correct text for every part of the organisation.

Manuals' narrative structures are in the middle of a report and a memorandum, nevertheless they have elements that draw lines into the future of possible situations of the organisation.

More on business development.

The frame: Mexican Petroleum is a state owned enterprise dedicated to produce and commercialise oil and its derivatives.

Scientific methods might be seen as good help to quantify the risk and return of certain options for management, but as seen in previous chapters the total epistemology of business developments and management science lies thoroughly on the scientific approach.

In the case of Blackett the generation of images in a visual literature style was sometimes the key to successful implementation of policies and generation of business. In fact, even sometimes drawings had to be made in the presentation of formal documents.

The general epistemology followed in several cases in PEMEX followed a hermeneutical process similar to the one proposed by Fisher in his statistical methods for research workers, where information and its management was considered of the utmost importance.

Experiments or virtual experiments were conducted as generation of information and proper implementation was achieved after a careful mathematical modelling.

Of course the implementation had a lot of further regulations and adaptations due to many unknown and volatile factors in the market, law and pricing schemes.

The general kind of problems that one could face when establishing a strategy for business development in Latin America could lead to a thorough analysis in a more stable environment. For example, at the time of evaluating risk, the political, economic and natural risks are of the utmost importance. However, changes in quality or conditions of suppliers, changes in consumer needs and the entrance of substitute goods in the market were normal risk included in the evaluation. Labour, credit and behavioural risks were also on top of the analysis.

There is a common factor in the epistemology of management science linked to strategy: the mathematics applied to quantify the circumstances and problems are mostly linear, for example linear models, such as Fisher Discriminant Algorithm (FDA), linear programming, SIMPLEX Method, differential coefficients, and even some network analysis are commonly used. Planning obeys to stochastic processes and are all linear approximations, Blackett exactly as in the case of Fisher and Blackett, the only difference is the virtuality of the experiments.

Business systems and applied mathematics:

A business system is basically the set and organisation of resources by means of which an enterprise makes profits. In classical analysis, organisational systems refers to the way in which people and resources get together and accomplish to achieve business.

Members, structure, processes and culture are main parts of the business system. These parts are indeed connected in a strategic, or I would call a mathematical way. The optimisation of resources obeys to quantification of results, and strategic renewals depend on the way in which the structural behaviour is planned.

Simulation Blackett's Style.

Due to the randomeness not only of oil markets, level of demand of products, and interests rates, business simulation is done via probability generally applying statistical methods normal distribution techniques.

Of course, what we call Monte Carlo simulation is based on statistical techniques, and experimental simulation obeys to both Fisher's design of experiments and Blackett's application of Poisson Distribution to problems involving events in time intervals. In fact in [Anderson 1994, 557], we have documented a case in Mexico in which this kind of simulation was applied.

Sensitivity analysis for taking decisions to develop business, also were in line with Blackett's methods illustrated also in his work on cosmic radiation.

Analysis was performed by a specialised department of business planning. In this case, the evaluation and development of business schemes was a complex and multidisciplinary one, as it involved technological issues, such as the knowledge of pipelines, cryogenic tanks and refinery processes, law issues as the properties of PEMEX (depending on the products) were state owned. The economic, and financial evaluations were of the utmost importance and performed simultaneously by several teams and discussed in consensual meetings.

A main characteristic in PEMEX was the division of tasks in strategic business units. Planning was centralised, so integration and coordination of the organisation was difficult in terms of timing. Once a project had been evaluated by one business unit, then it passed to the next one. In the case of military organisation the structure was similar. Blackett's circus is a model of strategic multidisciplinary team.

More than going in the details of the methods and techniques for evaluation, the main idea was to present the results in such a way that the recommendation was successful. Sometimes, the projects could not go ahead due to some political issues and other times simply because other departments in the overall evaluation were affected in a greater way.

In this sense, Latin American business had the support of European and American science. In fact International firms such as McKinsey advise the most successful enterprises in Latin America. Their methodologies are those that we can find in textbooks, however, the most important issue lies beside the methodology: the epistemology that is constructed on the line of a specific project.

The methods used for the evaluation of projects were standard and documented in many specialised literature. But the most important issue is that it was not the methodology, but the way in which communication of results was made, that companies like PEMEX made decisions on business and capital investment projects in general.

As we could see from previous chapters, the methodology followed by Blackett, in the case of Britain, to evaluate situations and explain them in a scientific way (quantitative) was not the state of the art methodology and was not at all the top scientific knowledge these scientists had in their base. The sociological argument behind this fact is that militaries understood better statistical arguments than further mathematical analysis. Simply by putting the data in a proper way, they executive levels in the organisation could realise about important variables on certain operation and could improve tactics and strategies based on that.

As exposed in the first chapters, Blackett had a main task in his job as a military advisor, this was communication; the literature he developed was aimed for the level of understanding of the military at that moment, and the analysis, working papers and notes found were only based on general statistics and not on any heavier mathematical artillery.

A general case is exposed in [De witt, 2004, 75], as told by Mr. Ohmae, head of McKinsey office in Japan, "As a consultant I have had the opportunity to work with many large Japanese companies. Among them are many companies whose success you would say be the result of superb strategies. But when you look more closely you discover a paradox. They have no big planning staffs, no elaborate, gold platted strategic planning processes. Some of them are painfully handicapped by the lack of resources –people, money and technology- that seemingly would be needed to implement an ambitious strategy".

In the case of many Latin American and British companies this is not the exception.

Latin America has had its own cases of developing applied science, an outstanding case is the one of Julio Rey Pastor, who developed excellent literature on books like "Los Problemas Lineales de la Fisica" in which he exposes the cases such as Hilbert Spaces and geometric problems associated, variational calculus and linear problems and other applications of theorems and equations to many cases of physics.

However, methodologies and scientific ideas were developed in Latin America, as [Canaparo 2001], states about the case of "Redes" publication following the ideas and formats of "Nature", Latin American incursion of scientists in the field of business their evaluation and development was also following models like Blackett's, Fisher's or Kantorovich's.

We can see how in the case of PEMEX the evaluation of projects mimicked Blackett's case in many ways. For example the types of analysis delivered for the evaluation of projects was of the highest standards, however, the most important fact was their communication to higher officers in the government and executive levels in the same organisation.

As in [Anderson 1997], eight chapters of the book on "management science" are dedicated to linear programming and its applications, work developed mainly by Kantorovich, Dantzig and others in the last decades of the 20th century.

For the strategic part, we can see how epistemological similarities point to the original work of Blackett. For example, [Dess, 1993, 93] states that an important step to achieve a good business strategy is to identify success factors. In this sense he had analysed in previous chapters [54] a table measuring the impact of several variables or fields of impact towards a certain project, very much in line with the developments of Blackett in the military organisation in Britain.

In [De wit, 2004] we can see how the concept of an actor as the executive part in a system is taken from what we could read on

Latour's Science in Action; in fact the organisational activities in the strategic plans as outlines in chapter 3 are similar to Blackett's planning in the military organisation.

Moreover, the term strategy comes precisely from military activities, [De witt, 2004, 30]. In a way De witt follows a constructivist approach, very much in line with our current epistemological frame.

His steps on identifying, diagnosing realizing, and conceiving are worthy of epistemic analysis by themselves, as put by De witt, they interact in a hermeneutical way towards what he calls knowledge.

Planning business in Mexico at this level implied an awareness of many factors. Projects began a deep analysis of existing data, recollection of data was not a big task due to many existing databases and hired publications they had in the general offices, although further data had to be searched. Basic planning steps were respected, however further planning adjustments had to be performed. Problem-solving strategies had to be implemented, for example, the conduction of experiments in a virtual way, which were less costly than those performed by the same Blackett with the military.

After the collection of data, information was generated and used immediately for decision-making purposes.

The strategic chain: define, diagnose, design decide was modified in many ways due to many factors, such as the availability of information, and other important decisions that had to be made often by third parties.

The formula of making decisions through discussion, collage and improvisation was more the state of affairs in the Mexican environment than a perfect scientific method such as exposed by Blackett.

Complex models using differential topology models were suggested to evaluate projects, however the simplest methods proved to be the most accurate and successful. Given the time constraint for the negotiations, held more pressure due to situations difficult to handle, such as the issue was not clear nor the data reliable, many elements had to be combined and sometimes communication across boundaries of planning departments were not fluent. As one can imagine, all the situations were novel and sometimes confusing and complicated specifications were often presented by business partners.

In the Mexican case scientific method did not play a role as significant in the origins of applied mathematics to these matters. Instead, imagining new solutions and to or what experts like [De witt, 2004] says option generation and selection were then first epistemic step in the Latin American case.

In an epistemological sense, the formation of groups and the utilization of what we know as operational research techniques, is more common than we thought.

Conclusion

In this work I have analyzed the work how the generation of mathematical knowledge is achieved in practical cases, and also have pointed crucial factors for its successful applications. The main applications, as we have seen are in the field of strategic planning, being the main features not only the technical mathematics, but also the communicative. The emergence of Operational Research as a tool to evaluate and develop strategy in applied cases was a clear example, above all establishing a direct connection between the methods used in cosmic radiation studies and in Management Science, I have also made some remarks on the construction of applied mathematics by a peculiar use of language.

Operational Research developed ideas like backing decisions on scientific grounds and contributed to establish a tradition of cooperation between scientists and executives in organizations. These scientists can work closely and effectively in problems of a strictly non-scientific nature. This contributed to place physicists and mathematicians in high executive levels of non-academic organizations, such as the military, industry, and the national economy.

The key issue for any analyst, be financial, or strategic, is the way of looking to applied problems, for example the case of war he has to represent them and interpret them, in other words, he has to create a reality principle via the scientific discourse. This discourse as a cognitive issue is a major breakthrough in the field of applied mathematics. It is important to note that having established the epistemological limits of language in the scientific contexts is enough to understand the reach of the discipline.

In the case of the design of a technique for experiments, and other related statistical concepts, the epistemology obeyed also to a general discourse in which the generation of an expecting horizon in terms of P.Ricoeur [1983–1985] was a main objective. This construction was extrapolated via statistical methods, which were used as shortcuts of an extensive discourse that uttered knowledge in the form of a theory. The importance of this discourse (technical and colloquial) was raised and became a guiding principle for research.

This new rhetoric for strategy had a unique feature, which was the duality of "theory and methodology" which transformed the subject (of designing experiments) in an epistemology in itself. In other words, applied mathematics can be interpreted as an epistemic ground to launch specific knowledge coming from different areas (it is a *sui generis* way of knowing); in fact, the theory is also the methodology in the case of experiments; a clear example is the experimental trials nowadays, this trials have a proper narrative of experiments (reports and results) which developed knowledge in a specific way (the use of statistics to mention one).

Finally concerning the economic and computational applications I have attempted to show how the way in which a rhetoric finally ended in a hermeneutical process that was connected with the action in the "real world" (methods and decision taking in big organizations).

A common and important issue bound all the cases here exposed: legitimization. This legitimization is based on scientific grounds. For example the design of an experiment had, as a main purpose, the legitimization and validity of the results and conclusions of it. In Operational Research, the purpose of the discourse was to support some measures taken at executive levels (making the decision legal), and not creating a model to solve or optimize some process; this kind of approach was typically used in other areas (in physics as a given equation in a cosmic radiation problem). The generation of an economic model in a world where economics also is considered politics could only be achieved mixing epistemologically pure mathematics and real necessities; in this last case, the root of legitimization (lex=law) is the most specific.

Another common issue is the fact that all the cases worked in multidimensional spaces; the case of Operational Research is clear, as its definition as strategic planning is dealing directly with functions in multidimensional spaces. Fisher's correlation coefficient is taken as the inner product, with multidimensional vectors. The case of Linear Programming is absolutely clear, as the general linear programming problem is the maximization or minimization of a certain linear multidimensional function subjected to other multidimensional functions.

The general narrative understood as knowledge advised us to go deeper in our work in an abstract study of texts. These texts were immersed in an epistemic frame, which we considered to be the constructivist; due to the nature of the discipline (applied mathematics). Other philosophical and historiographic tools were also applied, but all converging to define a general epistemological ground, as proposed in the thesis of the book.

Finally, the generation of objects/concepts in the classic historiographic sense generated a territory, specially in international cases like Latin America, a space in which events (including cognitive events) interacted with other elements in the formation of a body of science. These concepts usually do not follow a time-based logic, but a spatial one, such as modern strategies to evaluate strategic planning.

In this way, the best way to read the scientific discourse in the business context, in our opinion, is through the resource of a map, developing a cartographic image and study (in the cognitive sense). In it, the epistemic platform serves as a compass to navigate, and that is why we found that it was so important to delimit the reaches and constraints of cognitive issues in each of the disciplines considered.

Bibliography

Primary Sources.

Bodleian library archives
Bernal Papers
Darlington Papers
Myers Papers

Cambridge University library
Bernal Papers
Fisher Papers

Imperial College archives
Tizard Papers

Imperial War Museum
Tizard Papers

Institute of Historical Research, Senate House, University of London
Military Papers

King's College archives
Liddell-Hart Papers

Public Record Office
Admiralty papers
Air Ministry papers
Cabinet Office papers
Home Office papers
Prime Minister Office papers
Treasury papers
Chancellor of Exchequers office papers

Royal Society archives
Blackett Papers

Secondary Sources.

Air Ministry, *The Origins and Development of Operational Research in the Royal Air force,* Her Majesty's stationery office, London, 1963.

Anderson E., *The management of manufacturing: Models and Analysis*, Addison-Wesley, Workingham, 1994.

Andrada M., *Instruccion que a V. Magestad se da para mandar fortificar el mar Oceano y defender de todos los contrarios Piratas, ansi Franceses, como Ingleses, en todas las navegaciones de su Real Corona, dentro de los Tropicos*, 1590.

Barthes R., *Le degré zéro de l'écriture,* Seuil, Paris, 1953.

——, *L'aventure semiologique,* Seuil, Paris, 1985.

Bazerman C., *Textual Dynamics of the Professions: Historical and Contemporary Studies of writing in Professional Communities,* University of Wisconsin Press, Madison, 1991.

Beadle, M. E., "Television and Argentine perceptions of the United States", Paper presented at the Conference of the Americas, Mexico City, Mexico, 1991.

Bennett, J.H., *Natural Selection, Heredity and Eugenics,* Clarendon Press, Oxford, 1983.

Birkhoff G., *Dynamical Systems,* American Mathematical Society, Providence, 1927.

Blackett P. M. S., *La radiation cosmique,* Hermann & Co., Paris, 1935.

——, *Cosmic Rays: The Haley Lecture,* Clarendon Press, Oxford, 1936.

——, "Evan James Williams, 1903–1945", *Obituary notes of the Fellows of the Royal Society* Vol 45, 387, 1947.

——, "Operational Research", *Advancement of Science,* 5:17, 26–38., 1948.

——, "Operational Research", *Operational Research Quarterly,* Vol. 1, pp. 3–6, 1950.

——, "Operational Research: Recollections of Problems Studied, 1940–45", *Brassey's Annual,* 1953.

——, "Tizard and the Science of War", Tizard Memorial Lecture delivered on February 11 the Institute of Strategic Studies, reprinted in *Nature* Vol. 185 No. 4714 pp. 647–653, March 5, 1960.

——, *Studies of War: nuclear and conventional,* Oliver & Boyd, London, (1962).

Bloor D., *Knowledge and Social Imagery,* Chicago University Press, Chicago, 1976.

Blumenberg H., *Die Genesis der kopernikanischen Welt,* Suhrkamp, Frankfurt, 1975.

Brentjes S., *Linear optimization, Companion Encyclopaedia of the history and philosophy of the mathematical sciences,* Routledge, London, 1994.

Calvino I., *Lezioni Americane: Sei proposte per il prossimo millennio,* Oscar Mondadori, Milano, 1993.

Canaparo C., *Imaginacion mapas y escritura*, University of Exeter, Exeter, 2000.

——, *The manufacture of an Author,* Centre for Latin American Studies, King's College London, London, 2000a.

——, *El perlonghear,* University of Exeter, Exeter, 2001.

——, *Un mundo modernista para la cultura rioplatense,* La protesta ediciones, Buenos Aires, 2002.

Charnes A. / Cooper W.W., "On Some Works of Kantorovich, Koopmans and Others", *Management Science,* Vol. 8 No. 3, 246–263, 1962.

Clausewitz von K., *On War,* Princeton University Press, New Jersey, 1976.

Courant-Hilbert, *Methods of mathematical physicis*, Interscience Publishers, New York, 1953.

Dando M.R./ Sharp R. G., "Operational Research in the U.K. in 1977: The Causes and Consequences of a Myth?", *The Journal of the Operational Research Society,* Vol. 29 No. 10, pp. 939–949, 1978.

Dantzig G.B., *Linear Programming and Extensions.* Princeton University Press, Princeton, New Jersey, 1963.

D*eleuze* G., *Cinema I,* Minuit, Paris, 1983.

——, *Cinéma* ii, Minuit, Paris, 1985.

Deleuze G./Guattari F., *Qu'est-ce que la philosophie?*, Minuit, París, 1991

Derrida J., *L'ecriture et la difference,* Éditions du Seuil, Paris, 1967.

——, *De la Grammatologie,* Minuit, Paris, 1967a.

——, *Positions,* Collection Critique Minuit, Paris, 1972.

Dess G., *Strategic Management*, Mc Graw Hill, New York, 1993.

De Wit B., *Strategy--process, content, context : an international perspective,* Thompson, London.

Eco U., *A theory of semiotics,* Indiana University Press, Indianapolis, 1976.

——, *Lector in fabula,* Bompiani, Milano, (1979).

——, *Interpretazione e sovrainterpretazione,* Bompiani, Milano, (1995).

Eilon S., "How Scientific is Operational Research", *Omega,* Vol.3, pp.1–8, 1975.

Feyerabend P., *Philosophical Papers Volume I,* Cambridge University Press, Cambridge Mass, 1981.

——, *Philosophical Papers Volume II,* Cambridge University Press, Cambridge Mass, 1981a.

Feinberg, S., *R.A. Fisher: An appreciation,* Springer-Verlag, New York, 1980.

Fisher, R. A., *Statistical methods for research workers,* Oliver and Boyd, London, 1925.

——, *The Genetic Theory of Natural Selection,* Oliver and Boyd, London, 1930.

——, *Statistical methods for research workers,* 4th edition, Oliver and Boyd, London, 1932.

——, *The design of experiments,* Oliver and Boyd, London, 1935.

——, *Collected Papers,* Vol II: 1925–1931, University of Adelaide, Adelaide, 1972.

Fisher, J., *R.A. Fisher: the life of a scientist,* John Wiley and sons, New York, 1978.

Forder R.A., *Operational analysis in defence—a retrospect, defence evaluation and research agency,* http://home.freeuk.com/forder/retrospect/retrospect_text.htm (2000).

Fortun M., "Scientists and the legacy of World War II: the case of operational research (OR)", *Social Studies of Science* vol. 23, 595–642, 1993.

Fuller, S.: *Thomas Kuhn. A Philosophical History for Our Times,* Chicago University Press, Chicago, 2000.

Galison P., *Image and Logic,* The University of Chicago Press, Chicago, 1997.

Gardner R., "L.V. Kantorovich: the price implications of optimal Planning", *Journal of Economic Literature,* Vo. 28, No. 2, 638–648, 1990.

Genette, G., "Discours du récit", in *Figures III*: Paris: Seuil, 1972.

Glasersfeld E. von, *Radical Constructivism. A way of Knowing and Learning,* The Falmer Press, London, 1995.

Grattan-Guinness I., "'A new type of question': On the prehistory of linear and non-linear programming, 1770–1940", *The History of Modern Mathematics,* 3, Academic Press Inc., London, 1994.

Gross, A.: *The rhetoric of science,* Harvard University Press, Cambridge (Mass), 1996.

Hadamar J., *The psychology of invention in the mathematical field,* Dover Publications, New York, 1954.

Hand D. J., "Statistics and the theory of measurement, (with discussion)". *Journal of the Royal Statistical Society,* Series A, 159, 445–492, 1996.

——, *Principles of data mining,* MIT Press, Cambridge Mass, 2001.

Heidegger M., *Being and Time,* Basil Blackwell, Oxford, 1995.

Isbell J.R., "On an Industrial Programming Problem of Kantorovich", *Management Science,* Vol. 8 No. 1, 13–17, 1961.

Jones R. V., "Lord Cherwell's Judgement in World War II", *The Oxford Magazine,* Oxford, 9 May, 1963.

Johansen L., "Soviet Mathematical Economics", *The Ecnomic Journal,* Vol. 76, No. 303, 593–601, 1966.

Kaijser A., *From Operations Research to futures studies: The establishment, diffusion, and transformation of the systems approach in Sweden, 1945–1980,* 2000.

Kantorovich L.V., "A new Method of Solving of Some Clases of Extremal Problems", *Compte Rendus de la Acdemie des Science de l'URSS,* Vol. XXVIII, No. 3, 211–214, 1940.

——, "On the Translocation of Masses", *Compte Rendus de la Acdemie des Science de l'URSS,* Vol. XXXVII, No. 7 8, 199–201, 1942.

——, "Mathematical methods of Organizing and Planning Production", *Management Science,* Vol. 6, 366–422, 1960.

——, "Mathematics in Economics: Achievements, Difficulties, Perspectives", *The American Ecnomic Review,* Vol. 79 No. 6, 18–22, 1989.

Kirby M.W., "The air defence of Great Briatin 1920–1940: an operational research perspective", *Journal of operational research society,* Vol. 48, 555–568, 1997.

——, *Operational Research in war and peace,* Imperial College Press, London, (2003)

Kircher P., "Translator's Note on 'Mathematical Methods in Economics'", *Management Science,* Vol. 7 No. 4, 335–336, 1961.

Kittel C., "The Nature and Development of Operations Research", *Science Magazine,* 105, pp.150–153, 1947.

Knobloch E., "Mathematics at the Berlin Technische Hoschschule / Technische Universität. Social, Institutional, and Scientific Aspects", *The History of Modern Mathematics,* Academic Press, New York, 1989.

Koopmans T.C., "A Note about Kantorovich's Paper, 'Mathematical Methods of Organizing and Planning Production'", *Management Science,* Vol. 6, No. 4, 363–365, 1960.

——, "On the Evaluation of Kantorovich's Work of 1939", *Management Science,* Vol. 8 No. 3, 264–265, 1962.

Larnder H., "The origin of operational Research", *Operational research '78,* North-Holland, 1978.

Latour B., *Science in Action: how to follow scientists and engineers through society,* Harvard University Press, Cambridge Mass, 1987.

——, *Petites Lecons de sociologie des sciences,* La Découverte, Paris, 1993.

Lindbeck A., *Nobel Lectures, Economics 1969–1980,* World Scientific Publishing Co. Singapore, http://nobelprize.org/economics/laureates/1975/kantorovich-autobio.html, 1992.

Lovell B., *P.M.S. Blackett, a biographical memoir,* The Royal Society John Wright & Sons LTD, London, 1976.

Luvaas J., *Frederick the Great on the Art of War,* Free Press, New York, 1966.

Lyotard J. F., *La Condition postmoderne: rapport sur le savoir.* Collection "Critique." Paris: Minuit, 1979.

——, *Phenomenology,* State University of new York Press, New York, 1986.

——, *Le postmodern expliqué aux infants,* Galilée, Paris, 1988.

Machiavelli N., *The Art of War,* Da Capo Press, New York, 1965.

Martin, J. & Chaney, L., "Determination of content for a collegiate course in intercultural business communication by three delphi panels", *Journal of Business Communication*, 29 (3), 267-283, 1992.

Maturana H., *La realidad: obejtiva o construida?*, Anthropos / Universidad Iberoamericana / ITESO, Barcelona, 1995.

McArthur C. W., *Operations Analysis in the U.S. Army Eight Air Force in World War II,* American Mathematical Society, Providence, 1990.

McCloskey J. F., "US operations research in World War II", *Operations Research* 35, 910–925, 1987.

——, "The beginnings of operations research: 1934–1941", *Operations Research* 35, 148, 1987a.

Merton R., *The sociology of science,* Chicago University Press, Chicago, 1973.

Mindell D. A., "Automation's finest hour: Radar and system integration in World War II", *Systems, Experts and Computers,* Hughes and Hughes, MIT Press, Boston, 2000.

——, "Topology and physics—a historical essay", *History of Topology,* Elsevier Science, London, 1999.

Nye M. J., "Temptations of theory, strategies of evidence: P.M.S. Blackett and the earth's magnetism, 1947–1952", *British Journal of the History of Science,* Vol. 32, 69–92, 1999.

——, *Blackett: Physics War and Politics in the Twentieth Century,* Harvard University Press, Cambridge Mass., 2004.

Operational Research Quarterly, Vol 13:3, 282, 1962.

Ortega y Gaset J., *The revolt of the masses,* W.W. Norton, New York, 1993.

Owens L., "Mathematicians at war: Warren Weaver and the applied mathematics panel, 1942–1945", *The History of Modern Mathematics,* Academic Press, Inc, New York, (1989).

Pasley C.W., *Military Policy and the Institutions of the British Empire,* William Clowes and Sons Limited, London, 1914.

Pearce M., *British political history 1867–1990,* Routledge, London, 1992.

Peirce C. S., *Writings of Charles S. Peirce,* Volume 2 1867–1871, Indiana University Press, Indianapolis, 1986.

——, *Writings of Charles S. Peirce,* Volume 3 1872–1878, Indiana University Press, 1986a.

Piaget J., *Six psychological studies,* Vintage, New York, 1967.

——, *L'épistémologie génétique,* PUF, Paris, 1970.

Pile F., *Ack Ack,* Marshall Simpkin, London, 1945.

Plata, S., "A note on Fisher's Correlation Coefficient", *Applied Mathematics Letters,* Ref. AML4650, Elsevier, Washington, 2005.

Poincare H., *Collected Works,* Volume VI, Analysis Situs, pp.1 86, publiees sous les auspices du Ministere de l'instruction publique, Paris, 1895–1928.

——, *Science and Hypothesis,* Dover Publications, London, 1903.

——, *Science and Method,* translated from French by Francis Maitland, Dover Publications, London, 1906.

——, *Last Essays,* Dover, New York, 1913.

Putnam H., *Philosophical Papers: Mathematics matter and method,* Cambridge University Press, Cambridge Mass, 1975.

Rau E. P., "The adoption of Operations Research in the United States during World War II", *Systems, Experts and Computers,* Hughes and Hughes, MIT Press, Mass, 2000.

——, "Technological systems, expertise, and policy making: The British origins of Operational Research", *Technologies of Power,* The MIT Press, Mass, 2001.

Ravindran A., *Operations research: Principles and Practice,* John Wiley & Sons, New York, 1976.

Ricoeur, P., *Temps et récit I,* Editions du Seuil, Paris, 1983.

——, *Temps et récit II,* Editions du Seuil, Paris, 1984.

——, *Temps et récit III,* Editions du Seuil, Paris, 1985.

Rider R., "Operations Research and game theory", *History of political economy* 24, suppl. S225–s239, 1992.

——, *Operational Research, Companion Encyclopaedia of the history and philosophy of mathematical sciences,* Routledge, London, 1994.

Rosenhead J., "Operational research at the Crossroads", *Journal of Operational Research Society,* 40, 3–28, 1989.

Rutherford / Chadwick / Ellis, *Radiations from radioactive Substances,* Macmillan Pub. Company, New York, 1930.

Schubring G., *Pure and applied mathematics in divergent institutional settings in Germany: The role and impact of Felix Klein, The History of Modern Mathematics,* Academic Press, Inc, 1989.

——, "Klein and German mathematics", *The History of Modern Mathematics,* Academic Press, Inc, 1989a.

Segal J., *Le Zéro et le Un – Histoire de la notion scientifique d'information,* Edition Syllepse, Paris, 2003.

Sheynin O. B., *On the Mathematical treatment of observation by L. Euler,* 1971.

Sinai Y., *Russian Mathematicians in the 20th century,* Princeton University Press, Princeton, 2003.

Siroyezhin I., "Oprations Research in the USSR as Education and Research Work", *Management Science,* Vol 11, No. 5, 593–601, 1965.

Steiner G., *Real presences,* Faber and Faber, London, 1989.

Saussure F. de, *Course of general linguistics,* McGraw-Hill, London, 1959.

Tizard H., *A scientist in and out of the civil service,* Haldane Memorial Lecture, Birkbeck College, 1955.

Tobies R., "On the contribution of mathematica*l societies to promoting application of mathematics in Germany", The History of Modern Mathematic,* Academic Press, Inc., New York, 1989.

Virilio P., *War and Cinema: the logistics of perception,* Verso, London, 1984.

Vucinich A., *Empire of Knowledge the academy of science of the USSR (1917–1970),* University of California Press, Berkeley, 1984.

Waddington C.H., *O.R. in World War 2,* Elek Science, London, 1973.

——, "P.M.S. Blackett: an appreciation", *Operational Research Quarterly* 25:4, editorial I, 1974.

Ward B., "Kantorovich on Economic Calculation", *The Journal of Political Economy,* Vol. 68 No. 6, 545–556, 1960.

Hispanic Studies: Culture and Ideas

Edited by
Claudio Canaparo

This series aims to publish studies in the arts, humanities and social sciences, the main focus of which is the Hispanic World. The series invites proposals with interdisciplinary approaches to Hispanic culture in fields such as history of concepts and ideas, sociology of culture, the evolution of visual arts, the critique of literature, and uses of historiography. It is not confined to a particular historical period.

Monographs as well as collected papers are welcome. Languages of publication are English, Spanish and Spanish-American.

Those interested in contributing to the series are invited to write with either the synopsis of a subject already in typescript or with a detailed project outline to either Dr Claudio Canaparo, Centre for Latin American Studies, School of Arts, Languages and Literatures, University of Exeter, Exeter EX4 4QH, UK c.canaparo@exeter.ac.uk, or to Alexis Kirschbaum, Peter Lang Publishing, Evenlode Court, Main Road, Long Hanborough, Witney, Oxfordshire OX29 8SZ, UK, a.kirschbaum@peterlang.com.

Vol. 1 Antonio Sanchez
Postmodern Spain. A Cultural Analysis of 1980s–1990s Spanish Culture. 220 pages. 2007.
ISBN 978-3-03910-914-2

Vol. 2 Geneviéve Fabry and Claudio Canaparo (eds)
The Enigma of Reality. Towards an Oblique Expression of Reality in 20th-Century Narrative Literature. *Forthcoming.*
ISBN 978-3-03910-893-0

Vol. 3 William Rowlandson
Reading Lezama's *Paradiso*. 290 pages. 2007.
ISBN 978-3-03910-751-3

Vol. 4 Claudio Canaparo, Fernanda Peñaloza and Jason Wilson (eds)
Patagonia. Myths and Realities. *Forthcoming.*
ISBN 978-3-03910-917-3

Vol. 5 Xon de Ros
Primitivismo y Modernismo. María Blanchard y los escritores de 1927.
238 pages. 2007.
ISBN 978-3-03910-937-1

Vol. 6 Sergio Plata
Visions of Applied Mathematics. Strategy and Knowledge.
284 pages. 2007.
ISBN 978-3-03910-923-4

Vol. 7 Annick Louis
Borges y el Fascismo. *Forthcoming.*
ISBN 978-3-03911-005-6

Vol. 8 Helen Oakley
From Revolution to Migration. A Study of Contemporary Cuban and Cuban American Crime Fiction. *Forthcoming.*
ISBN 978-3-03911-021-6

Vol. 9 Thea Pitman
The Chronotrope of the Net. The Postmodern and the Postcolonial in Contemporary Mexican Travel-Writing. *Forthcoming.*
ISBN 978-3-03911-020-9

Vol. 10 Francisco J. Borge
A New World for a New Nation. The Promotion of America in Early Modern England. 240 pages. 2007.
ISBN 978-3-03911-070-4

Vol. 11 Helena Buffery, Stuart Davis and Kirsty Hooper (eds)
Reading Iberia. Theory/History/Identity. 229 pages. 2007.
ISBN 978-3-03911-109-1

Vol. 12 Sheldon Penn
Writing and the Esoteric. Identity, Commitment and Aesthetics in Mexican Literature since the Revolution. *Forthcoming.*
ISBN 978-3-03911-101-5

Vol. 13 Angela Romero-Astvaldsson
La Obra Narrativa de David Viñas. Itinerario contextual y la nueva inflexión de *Prontuario* (1955) y *Claudia conversa* (1993). *Forthcoming.*
ISBN 978-3-03911-100-8

Vol. 14 Aaron Kahn
Siege, Conquest, and the Ambivalence of Imperial Discourse. Cervantes' La *Numancia* within the 'Lost Generation' of Spanish Drama. *Forthcoming.*
ISBN 978-3-03911-098-8

Vol. 15 Turid Hagene
Negotiating Love in Post-Revolutionary Nicaragua. The Role of Love in the Reproduction of Gender Asymmetry. *Forthcoming.*
ISBN 978-3-03911-011-7

Vol. 16 Yolanda Rodriguez
The Eighty Years War through Spanish Eyes. The Netherlands in Spanish Historical and Literary Texts (1548-1673). *Forthcoming.*
ISBN 978-3-03911-136-7

Vol. 17 Stanley Black (ed.)
Juan Goytisolo: Territories of Life and Writing. 202 pages. 2007.
ISBN 978-3-03911-324-8

Vol. 18 Maria T. Sanchez
The Problems of Literary Translation. A Study of the Theory and Practice of Translation from English into Spanish. *Forthcoming.*
ISBN 978-3-03911-326-2

Vol. 19 Matias Bruera
Meditations on Flavour. *Forthcoming.*
ISBN 978-3-03911-345-3